Sources and Studies in the History of Mathematics and Physical Sciences

Springer Science+Business Media, LLC

Sources and Studies in the History of Mathematics and Physical Sciences

K. Andersen
Brook Taylor's Work on Linear Perspective

J. Cannon/S. Dostrovsky
The Evolution of Dynamics: Vibration Theory from 1687 to 1742

B. Chandler/W. Magnus
The History of Combinatorial Group Theory

A.I. Dale
A History of Inverse Probability: From Thomas Bayes to Karl Pearson, Second Edition

A.I. Dale
Pierre-Simon Laplace, Philosophical Essay on Probabilities, Translated from the fifth French edition of 1825, with Notes by the Translator

P.J. Federico
Descartes on Polyhedra: A Study of the *De Solidorum Elementis*

B.R. Goldstein
The Astronomy of Levi ben Gerson (1288–1344)

H.H. Goldstine
A History of Numerical Analysis from the 16th through the 19th Century

H.H. Goldstine
A History of the Calculus of Variations from the 17th through the 19th Century

G. Graßhoff
The History of Ptolemy's Star Catalogue

A.W. Grootendorst
Jan de Witt's *Elementa Curvarum Linearum, Liber Primus*

T. Hawkins
The Emergence of the Theory of Lie Groups: An Essay in the History of Mathematics 1869–1926

A. Hermann, K. von Meyenn, V.F. Weisskopf (Eds.)
Wolfgang Pauli: Scientific Correspondence I: 1919–1929

C.C. Heyde/E. Seneta
I.J. Bienaymé: Statistical Theory Anticipated

J.P. Hogendijk
Ibn Al-Haytham's *Completion of the Conics*

A. Jones (Ed.)
Pappus of Alexandria, Book 7 of the *Collection*

Continued after Index

Jan de Witt's *Elementa Curvarum Linearum, Liber Primus*

Text, Translation, Introduction, and Commentary by
Albert W. Grootendorst

With the Help of Miente Bakker

With 150 Illustrations

Springer

Albert W. Grootendorst
Department of Mathematics (Emeritus)
Delft University
Aardbeistraat II
The Hague 2564 TW
The Netherlands

Sources and Studies Editor
Gerald J. Toomer
2800 South Ocean Boulevard, 21F
Boca Raton, FL 33432
USA

Library of Congress Cataloging-in-Publication Data
Witt, Johan de, 1625–1672.
[Elementa curvarum linearum-liber primus. English.]
Jan de Witt's "elementa curvarum: / text, translation, introduction, and commentary by Albert W. Grootendorst with the help of Miente Bakker.
p. cm. — (Sources and studies in the history of mathematics and physical sciences)
Includes bibliographical references and index.
ISBN 978-1-4612-7056-0 ISBN 978-1-4612-1238-6 (eBook)
DOI 10.1007/978-1-4612-1238-6
1. Geometry, Analytic—Early works to 1800. I. Grootendorst, A. W. II. Bakker, Miente. III. Title. IV. Series.
QA551.W7613 2000
516—dc21 99-053565

Printed on acid-free paper.

Production managed by Terry Kornak; manufacturing supervised by Joe Quatela.
Photocomposed copy prepared from the author's LaTeX files.

9 8 7 6 5 4 3 2 1

ISBN 978-1-4612-7056-0

piae memoriae
uxoris meae

Jan de Witt (1625–1672)

Adriaen Hanneman (1601–1671) pinxit.
Source: Museum Boijmans Van Beuningen, Rotterdam

Preface

The present book is a translation into English of *Elementa Curvarum Linearum–Liber Primus*, written in Latin, by the Dutch statesman and mathematician Jan de Witt (1625–1672). Together with its sequel, *Elementa Curvarum Linearum–Liber Secundus*, it constitutes the first text-book on Analytic Geometry, based on the ideas of Descartes, as laid down in his *Géométrie* of 1637. The first edition of de Witt's work appeared in 1659 and this translation is its first translation into English. For more details the reader is referred to the Introduction.

Apart from this translation and this introduction, the present work contains an extensive summary, annotations to the translation, and two appendices on the role of the conics in Greek mathematics. The translation has been made from the second edition, printed by the Blaeu Company in Amsterdam in 1684.

In 1997 the translator published a translation into Dutch of the same work, likewise supplied with an introduction, a summary, notes, and two appendices. This edition appeared as a publication of the Stichting Mathematisch Centrum Amsterdam. The present translation, however, is a direct translation of the Latin text. The rest of this work is an English version of the introduction, the summary, the notes, and the appendices, based on the Dutch original.

Of course, I am greatly indebted to several people who advised me during the process of translating. In the first place I thank my friend and colleague Dr. Miente Bakker whose critical mind and great skill in the field of text processing have substantially contributed to this publication.

A number of colleagues have read parts of the manuscript. They are: Dr. J.J. Schoorl, who checked the Introduction and to whom I owe a considerable increase of my knowledge of English grammar; Dr. Adrian Rice, who read the whole translation and the annotations to this translation; Dr. Anke van Zanten and her husband, Prof. A.J. van Zanten, who read the appendices; and Mrs. M. Lange and her husband, Dr. E. Lange, who checked the translation of de Witt's letter to van Schooten. Of course I am fully responsible for the remaining errors.

I would also like to thank Dr. J.P. Hogendijk, who introduced me to Dr. Rice. Many thanks also to Tobias Baanders, who drew the fine figures illustrating the text.

I am also grateful to the Management of Stichting Mathematisch Centrum in Amsterdam for giving their permission for this translation.

Finally I thank Springer-Verlag for the publication of this edition, especially Dr. Catriona Byrne and Dr. Thomas von Foerster, for their initiative and constant cooperation, the production editor Ms. Terry Kornak, and the copy editor Ms. Julia Brainin for the meticulous way they read and corrected the proofs.

The Hague, April 2000
A.W. Grootendorst

Contents

Albert Grootendorst (*1924) studied mathematics at Leiden University, where he obtained his Ph.D. for a dissertation on Theta-series in relative-quadratic number fields (1959); his supervisor was Professor H.D. Kloosterman. From 1952 to 1956 he was a teacher at a grammar school in the Hague. From 1956 to 1989 he was attached to Delft University of Technology as a member of the scientific staff, a lecturer, and a full professor, respectively. In cooperation with his colleague Professor B. Meulenbeld he wrote a textbook on analysis in three volumes. He also translated a number of seventeenth-century mathematical texts from Latin into Dutch and into English, and wrote a number of articles on the history of mathematics. With Professor W.C. Waterhouse and Dr. C. Greitzer he revised Clarke's translation of *Disquisitiones Arithmeticae* by C.F. Gauß. His main research interests are Greek mathematics and seventeenth-century mathematics.

1
Introduction

1. Almost every Dutch person knows that the famous Dutch statesman Jan de Witt (1625–1672) was Grand Pensionary of Holland and that he was murdered in a gruesome way by a mob, together with his two-years-older brother Cornelis. The murder took place on August 20, 1672, at *Het Groene Zoodje,* (now known as *de Plaats*) in the centre of the Hague. A statue on the spot serves to keep his memory alive.

De Witt is also known to many as the author of the *Waerdije van Lijfrenten Naer Proportie van Los-renten* (1671), a treatise on life annuities and one of the first applications of the theory of probability in economics, a book that was not "rediscovered" until two centuries later. Only very few, however, know him as a pure mathematician, who wrote the first textbook on Analytic Geometry, *Elementa Curvarum Linearum*, in two parts: *Liber Primus* and *Liber Secundus.*

The core of the present book is the original Latin text of de Witt's work and its translation into English, accompanied by explanatory notes. Two appendices, a summary, a bibliography and this introduction complete the book.

2. In 1637, the Leiden printer and bookseller Jan Maire published the first printed work written by René Descartes (1596–1650), entitled *Discours de la Méthode pour Bien Conduire Sa Raison et Chercher la Vérité dans les Sciences.* [*Treatise on the Method for Guiding One's Reason Well and for Seeking the Truth in the Sciences*]. The body of the work was followed by three appendices: *La Dioptrique, Les Météores,* and *La Géométrie.* The

title page does not mention the author's name, but everybody knew who it was.

The third appendix, *La Géométrie*, constituted a real revolution in mathematics: algebra (formally expressed in letters and symbols, not in words) was applied to geometry for the first time in history, at least in printed form. The study by Fermat (1601–1665) in the same field, entitled *Ad Locos Planos et Solidos Isagoge* (*Introduction to Plane and Solid Loci*), had circulated since 1629 in handwritten form and was not printed until 1679! It consisted of only eleven pages. For more details, the reader is referred to the introduction to the translation of *Liber Secundus.*

Thus, the way was cleared for the branch of mathematics that is now known as analytic geometry.

3. Among those immediately fascinated by Descartes' work was the Dutch mathematician Frans van Schooten Junior (1605–1660), who in 1645 succeeded his father, Frans van Schooten Senior (1581/82?–1645), as professor in the *Nederduitsche Mathematique en Wiskunde* at the Engineering School associated with the Leiden University. *Nederduitsche Mathematique* included, among other topics, practical aspects of mathematics, that were especially suited for surveyors and constructors of fortresses. The word *Wiskunde* is an original Dutch word for Mathematics; in this context it has the connotation of "pure" mathematics. The Dutch language is the only language with original mathematical terms that are not derived from Greek or Latin; we owe them to Simon Stevin (1548–1620).

Van Schooten Jr. immediately understood the paramount importance of the *Géométrie*, but he found it a very difficult book.

One might get the impression that Descartes purposely wrote in a style that was difficult to grasp. In his *Géométrie* he writes (page [301]):

> *Mais ie ne m"areste point a expliquer cecy plus en detail, a cause que ie vous osterois le plaisir de l"apprendre de vous mesme, & l"utilité de cultiver vostre esprit en vous y exerceant, qui est a mon avis la principale, qu"on puisse tirer de cete science.* [*But I shall not stop to explain this in more detail, because I should deprive you of the pleasure of mastering it yourself, and of the advantage of training your mind by working over it, which, in my opinion, is the principal benefit to be derived from this science.*]

As a result of a thorough study of the *Géométrie*, in which he involved many of his colleagues, van Schooten decided to translate this work from French into Latin, the Lingua Franca of the learned world in the seventeenth century, to make it accessible to many more mathematicians.

4. The first edition of this Latin translation, likewise published by Jan Maire in Leiden, appeared in 1649.
The title reads:

> *Geometria à Renato Des Cartes Anno 1637 Gallicè edita; postea autem unà cum Notis Florimondi de Beaune, (...) in Latinam linguam versa, & Commentariis illustrata, opera atque studio Francisci à Schooten, in Acad. Lugd. Batava Matheseos Professoris (...).* [*The Géométrie by René Descartes, published in 1637 in French, afterwards, however, together with the notes by Florimond de Beaune, (...) translated into Latin, and supplied with annotations, by the care and diligence of Frans van Schooten, Professor of Mathematics at Leiden University (...)*].

Indeed, van Schooten had–with great effort and diligence–composed an elaborate commentary, in which he entered into all kinds of detail and showed his wide knowledge of mathematics. As the title indicates, one of his colleagues, Florimond de Beaune (1601–1652), had also contributed to it.

Van Schooten remained fascinated by the *Géométrie*, and he actively propagated Descartes' ideas through the spoken and the written word. He passed his enthusiasm on to his pupils, who were all brilliant mathematicians. They included Christiaan Huygens (1629–1695), Hendrick van Heuraet (1634–1660?), Jan Hudde (1628–1704), and Jan de Witt, whose work is the central part of the present book.

5. Encouraged by the great success of his annotated translation, Frans van Schooten prepared a second edition, which appeared in two volumes: the first one in 1659 and the second in 1661, one year after his sudden death. The second volume was edited by his half-brother Pieter (1634–1671). Both volumes were greatly extended by contributions made by pupils and colleagues of van Schooten.

The first volume contained, apart from an emended text of the 1649 edition, two letters from Jan Hudde (who many times was made burgomaster of Amsterdam): one on the solution of algebraic equations and the other on a method for computing maxima and minima (of course without differentiation). Here he introduced his famous Hudde's Rule (see lit. [20]).

The second volume opens with an article by van Schooten himself, in which he explains Descartes' method in a general way. Then follows a treatise by de Beaune on the solution of algebraic equations, preceding the important and extensive contribution by Jan de Witt, the core of the present book: *Elementa Curvarum Linearum, Liber Primus* and *Liber Secundus.* This volume closes with an article, likewise by van Schooten, on solving geometric problems by means of algebra.

Before entering into de Witt's contribution in greater depth, a few words on the further history of van Schooten's *Geometria.* After the closing down of Elzevier's Company in 1681, a third, unaltered, edition appeared in 1683, published by the Blaeu Company, likewise in Amsterdam. In 1695 the fourth and last edition, enriched with annotations by Jacob (I) Bernoulli, was published by the Knoch Company in Frankfurt.

6. In effect, Jan de Witt had finished his contribution in 1649, but it did not reach its final form until 1656. In the meantime, he held several high offices that took up much of his time: Pensionary of Dordrecht in 1650, Grand Pensionary of Holland[1] in 1653. Originally, the latter was the office of the Government Attorney, i.e., the counsel for the defence of the rights of Holland. Later on, when the importance of Holland had grown, this position became equivalent to the position of today's Prime Minister combined with the position of the Minister for Foreign Affairs. Therefore it was an all-important matter that he could rely on the support of his teacher and friend Frans van Schooten Jr.

In a letter, dated October 8, 1658, de Witt thanked his friend for his help and added "*eene aenspraecke aan UE*", i.e., an exposition that he also used as a preface to the 1659 edition of his *Elementa Curvarum Linearum.*

In this exposition he explains his intentions. He was annoyed at the fact that the well-known conic sections–the parabola, the hyperbola and the ellipse–which are clearly plane curves, are generated by means of three-dimensional considerations, namely as the intersections of a plane with a cone or a cilinder, which are solids. The Greeks called these curves *τόποι πρὸς ἐπιφανείαις* (Latin: loci solidi), which means "curves on surfaces."

In de Witt's own words:

> *When I studied more carefully the books on the other curved lines–as far as they have been transmitted by the Ancients and explained by the later Mathematicians–I thought it absolutely against the natural order, which in mathematics one should respect as much as possible, to seek the origin of these curves in a solid and then to transfer them to the plane.*

Clearly, he was thinking of a kinematical treatment of the conics, made possible by the definition of a conic as the locus of a point that moves in the plane according to a certain mathematical prescription. van Schooten, too, set great store by this approach. In section 7 of this chapter we shall go into more detail.

7. A clear insight into the cooperation between Jan de Witt and his teacher and friend Frans van Schooten Jr., concerning the publishing of the *Elementa Curvarum Linearum,* can be gained from the twenty-one letters they exchanged in the period from February 1658 until March 4, 1660 (lit. [17]). As already mentioned, by about 1650, Jan de Witt had produced a first draft of his book. Probably he wrote this in his years as a lawyer in The Hague (1647–1650). We can infer this from a passage in the preface to

[1]Holland was the largest and most influential province of the Netherlands in the seventeenth century. In the late nineteenth century, it was split into the current provinces North Holland and South Holland. So in this context, Holland is not the informal name of the Netherlands.

the second edition of van Schooten's *Geometria* (1659). Here van Schooten writes about de Witt's contribution:

> *(...) quae jam ante decennium formata in conceptu huc usque delituerat.*
> [*(...) which, though prepared in outline already ten years before, has been concealed until now.*]

The cause of this delay is clear: de Witt's official duties asked so much of his time, that he did not have enough time left to prepare the *Elementa* for printing.

When, in 1658, van Schooten was allowed access to this work, he was very enthusiastic. He wrote to de Witt in February of that year:

> *I like it very much, as I think the ideas are very original and the way in which you express yourself is very correct and clear.*

In his opinion it would be of great advantage

> *to the geometry as well as to the algebra if Your Honour should be willing to bring it to light.*

In the same letter he offers his help in case of publication:

> *(...)I will undertake to remake all the calculations according to the method of Des Cartes, to check all carefully and to write it out meticulously and to add clear and careful drawings.*

For a good understanding: van Schooten is referring here to *Elementa Curvarum Linearum, Liber Secundus,* which is often referred to by the correspondents as the *Compositio Locorum Planorum et Solidorum.*

In the letter of October 8, 1658, Jan de Witt wrote that this *Compositio* must not be published without a preceding brief treatise on the nature and the properties of curved lines (*curvarum linearum*). This "brief treatise" is the present *Elementa Curvarum Linearum, Liber Primus.* It covers eighty-three pages in van Schooten's edition; the pièce de résistance, *Liber Secundus,* covers ninety-seven pages.

Jan de Witt had already finished this treatise on the nature and properties of curved lines and sent it, together with the aforementioned letter, to van Schooten. This letter, by the way, was intended to thank van Schooten for the revised and adapted *Compositio Locorum Planorum et Solidorum,* which he sent to de Witt in February 1658.

Thus, within eight months, van Schooten had completed his job of revising and adapting the first systematic and more or less complete textbook on analytic geometry.

In the same letter de Witt asked van Schooten if he would be so kind as to place references to the relevant theorems of Euclid's *Elementa* in the margin and add cross-references to preceding theorems in his own work.

In a letter dated May 7, 1659, van Schooten answers in the negative, but it is not quite clear if he is referring to the above-mentioned request or to the request to make changes in cross-references belonging to a modified part of the text. The latter request was made in a letter dated April 28, 1659.

The aforesaid letter of October 8, 1658, also served another purpose. In it, Jan de Witt strongly criticizes Apollonius, who–in his opinion–had made things much too complicated, and he shows that his own method is much simpler. He even compares the number of theorems involved!

Van Schooten did his job with great dedication. He not only spent much time and effort on the text, but he also drew the figures with great precision. Two letters (of April 7, 1659, and March 30, 1659) were devoted to the drafts of the figures. Two other letters–of February 10, 1660, and March 2, 1660–concerned the printed text. They listed eleven minor errata in the text and eight errata in the figures. In preparing my translation I found eleven additional small errors, which are listed at the end of section 9 of this introduction.

In all cases, van Schooten left the final decision to the responsible author Jan de Witt, who carefully read the proofs. He also kept in touch with the printer, Lodewijk Elzevier in Amsterdam. Somewhat bitterly he remarked that he only managed to get ten copies for himself and thirty copies for de Witt, "*which was the utmost I could agree on.*" Not a word about royalties!

During the progress of the work, Christiaan Huygens was asked for his opinion. His judgment was highly positive, not only as far as the book was concerned but also as regards de Witt as a mathematician. This can be inferred from a letter that Huygens wrote to John Wallis on June 6, 1659. At the end of this letter he touches upon this work of de Witt and adds:

> *Nullum aeque saeculum Geometrorum ferax fuisse arbitror, inter quos vir ille, si negotiis minus distringeretur vel principem locum obtinere posset.* [*In my opinion no century has been as abounding in mathematicians (as ours, transl.), amongst whom this man even might have reached the first place, if he had been less occupied by his official duties.*].

On July 22, 1659, van Schooten reported to de Witt that six sheets of his treatise had been printed, and on February 21, 1660, van Schooten's half-brother Pieter, who took care of the latest rectifications, brought six copies, printed *on large and fine paper*, to the author. The title page, however, gives the year 1659.

In a letter of March 2, 1660, Jan de Witt expressed his gratitude and mentioned a number of misprints; in the same letter he asked for a complete copy of van Schooten's commentary on Descartes' *Géométrie*, as soon as this should appear.

Van Schooten responded two days later; in his letter he promised that he would add a list of the misprints to de Witt's contribution; he also promised to send a copy of his Commentary on the *Géométrie*. This is van Schooten's last letter to Jan de Witt: on May 29, 1660, van Schooten suddenly died. The cause of his death is not known.

The preparation of the publication of van Schooten's Commentary (including the appendices written by some of his colleagues) was taken over by his half-brother Pieter, who sent the remaining sheets of the whole work, together with the errata in the *Elementa Curvarum Linearum*, to de Witt. The accompanying letter, dated December 17, 1660, is the last known letter in this correspondence; the answer to this letter, which (according to a note by Jan de Witt) was mailed the next day, has not survived.

As noted earlier, in 1683 a third, unchanged, edition of van Schooten's *Geometria* was published by Blaeu in Amsterdam, including de Witt's (likewise unchanged) contribution.

The fourth, and also the last, edition, expanded with remarks of Jacob (I) Bernoulli (1654–1705) and published in 1695 by Knoch in Frankfurt, also contained an unchanged edition of de Witt's contribution.

Finally, it must be mentioned that there also exists a copy of the *Elementa Curvarum Linearum* as an independent book. It is clearly a reprint of the relevant pages of van Schooten's *Geometria,* as is evident from the title page and the page numbering. A copy of this book is in the Library of the University of Utrecht (NL).

8. In this section we make some general remarks on the *Elementa Curvarum Linearum* and roughly sketch its position in the context of seventeenth-century mathematics.

i. As stated earlier, de Witt's work consists of two essentially different parts: *Liber Primus and Liber Secundus,* the first of which is a necessary introduction to the second.

This *Liber Primus* is written in a purely verbal style, without using our well-known mathematical symbols. In *Liber Secundus*, de Witt uses the "modern" notation, introduced by Descartes in his *Géométrie:* the first letters of the alphabet, $a, b, c, \cdots$ for known parameters and the last letters, u, v, w, x, y, z, for unknown parameters, and our familiar $+, -, \times, :$. Instead of $=$, he uses ꝏ and instead of $\pm$, he uses 8.

The most important difference between these two parts is of course the content. In the first part, de Witt gives a necessary geometric introduction to the conics; in the second part he applies Descartes' new analytic method to these conics.

ii. In *Liber Primus*, the conics are defined without the use of a cone or a cylinder, or, as Jan de Witt writes: *Absque ulla solidi consideratione*, i.e., Without considering any solid (Liber I, chapter iii, page [227]).

We already quoted de Witt's stabs at Apollonius and at the Ancients in general.

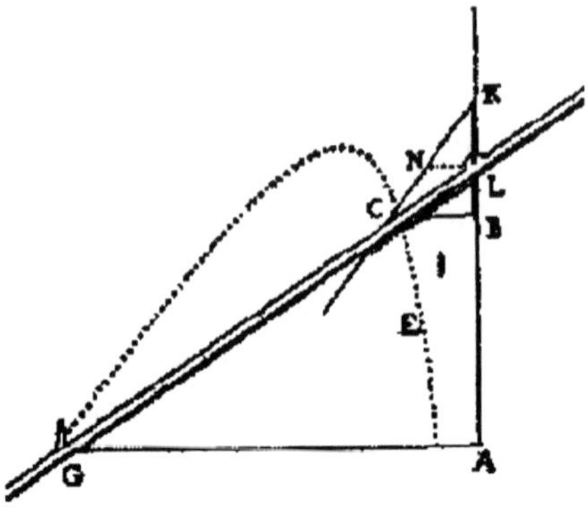

FIGURE 1.1. The generation of a hyperbola according to Descartes

Jan de Witt was not the first to draw a conic in the plane in a kinematical way according to a mathematical prescription. Archimedes (287–212), already, had given the well-known construction of an ellipse by means of two concentric circles and the excentric angle as the parameter (*de Conoidibus et Sphaeroidibus V*). de Witt gives this method (and a generalization) as a problem at the end of *Liber Primus.* de Witt's construction of an ellipse as the locus of a fixed point on a line segment, the endpoints of which slide along two fixed, intersecting straight lines, was previously known to Proclus (410–485 C.E.). de Witt's innovation, however, consisted in considering two arbitrary intersecting straight lines, instead of two mutually orthogonal lines. These two ways to generate an ellipse were also known to Claude Mydorge (1585–1647).

The so called "string construction" of an ellipse occurs previously in the work of Anthemius of Tralles (ca. 550 C.E.). This construction is based on the well-known focus property of an ellipse, already known to Apollonius (262–190 B.C.E.), and given in his *Konika* III, 52). In de Witt's *Liber Secundus*, we find this property as a problem on page [300] and the analogue for a hyperbola on page [291]. Isidorus of Milete (sixth century C.E.) gave the analogue for a parabola and Guidobaldo del Monte (1545–1607) gave the analogue for a hyperbola. Kepler (1571–1630) knew these three methods too.

For de Witt's kinematical construction of a parabola and a hyperbola, we refer to the summary and the translation. Here we only remark that the generation of a hyperbola by means of a rotating straight line and a sliding angle already occurs in the *Géométrie* of Descartes (page 320, see also Figure 1.1). According to Coolidge (lit. [6]), de Witt's generation of a parabola by means of a rotating angle and a sliding line is his own invention.

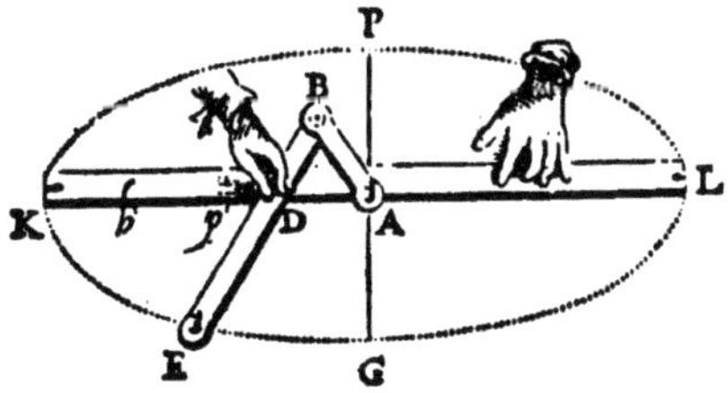

FIGURE 1.2. Kinematical construction of an ellipse according to van Schooten

iii. It is remarkable that de Witt finds no connection between these three conics, unlike Kepler, who regarded them as belonging to one family, the members of which can pass into each other continuously.

Notice, however, as de Witt himself pointed out, that his construction of a parabola can easily be modified in order to obtain a hyperbola merely by altering a certain angle (chapter IV). The ellipse in his considerations remains a distinct curve.

iv. The constructive, kinematical approach of the conics is consistent with the ideas of de Witt's teacher and friend, van Schooten Jr., who in 1656 or 1657 published his *Exercitationum Mathematicarum Libri Quinque* (later also published in Dutch: *Mathematische Oeffeningen* [1659–1660[). The fourth book is entitled *De Organica Conicarum Sectionum in Plano Descriptione Tractatus* (Dutch: *Tuych-werckelijcke Beschrijving van Kegelsneden*); it presents mechanical methods to describe the various conic sections.

In fact, this is a reprint of the geometrical part of a treatise from 1646, containing an appendix on the solution of the algebraic equations of the third degree.

In this work van Schooten Jr. takes great pains to construct the conics in a mechanical way, i.e., by means of various instruments, all based on a characteristic property of one of the conics. Here again we find the string construction of an ellipse and the analogous constructions of a parabola and a hyperbola. Figure 1.2 presents the sketch of an instrument by means of which van Schooten could describe an ellipse in yet another way. Here $KA = AL = AB + BE$ and $AB = BD$. When D moves along AK, the point E describes a quarter of an ellipse, which is not difficult to verify. These constructions also served a practical purpose: conics were useful in optics (in refraction as well as in reflection), in constructing sundials, and in the theory of perspective. Finally, notice, with regard to *Liber Primus,* that de Witt's way of generating a parabola and a hyperbola in chapter I can also be seen as a generation by means of projective pencils in the sense of the theory of Steiner (1769–1863) and Chasles (1793–1880), but de Witt does not appear to have had this in his mind. At this point, there is no need to go into more detail about *Liber Primus*. An elaborate survey of its contents is given in the summary preceding the translation.

v. After the kinematical introduction of the conics in *Liber Primus*, Jan de Witt operated in quite a different way in *Liber Secundus*. Here he applied the new analytical method, introduced by Descartes in his *Géométrie*, to the conics.

Unlike Descartes, who in fact only sketched his fundamental ideas, de Witt gave a nearly complete classification of the conics by means of the new method. Although I shall give a more extensive account of this method in the introduction to my translation of *Liber Secundus* (which is in preparation), let me summarize the essentials of this second book here.

In this book, Jan de Witt starts with an algebraic equation in two variables, x and y, of degree one or two, and he shows that these equations represent a straight line, a parabola, a hyperbola, an ellipse, or a circle.

To this end he selects a fixed line, a point A on it, and a fixed direction. Then all points in the plane can be determined by means of their abscissa x, measured from A along the fixed line, and their ordinate, measured from the endpoint of the abscissa in the chosen direction.

The variables and the coefficients in the equations are to be considered as line segments, and de Witt considers only points to the right of A and "above" the abscissa-axis chosen. We would say that he considers only the first quadrant, but this concept did not yet exist, as the y-axis was not introduced until the end of the seventeenth century, by C. Rabuel (1669–1728).

Next Jan de Witt constructs premeditatedly, on the basis of the coefficients occurring in the equation in question, a curve. Subsequently he selects an arbitrary point on this curve and shows, on the basis of the characteristic property ($\sigma\acute{\upsilon}\mu\pi\tau\omega\mu\alpha$) of the curve under consideration, that the abscissa and the ordinate of this point satisfy the equation under consideration.

However, this does not guarantee that there are no other points, besides the points on the curve under consideration, whose abscissa and ordinate satisfy the given equation!

There are several complications. As de Witt did not consider negative coefficients and as he knew no y-axis, he had to distinguish many more cases than we do, e.g.,:

$$
\begin{array}{lll}
y^2 = ax & \text{versus} & ay = x^2 \\
y^2 = ax + b^2 & \text{versus} & ay + b^2 = x^2 \\
y^2 = ax - b^2 & \text{versus} & ay - b^2 = x^2 \\
y^2 = -ax + b^2 & \text{versus} & b^2 - ay = x^2.
\end{array}
$$

It is evident that, in his opinion, $y^2 = -ay - b^2$ does not exist. In chapter I, he treats the linear equation. In chapter II he shows that the above equations represent parabolas and he reduces some more complicated equations to these "standard forms." He also shows that a quadratic equation in two variables represents a parabola if the quadratic terms form a complete square. de Witt concludes this chapter by showing that the locus of the points with the same distance to a fixed point and to a fixed straight line, is a parabola; here the concept of "focus" (umbilicus) is introduced.

In chapter III it is proved that the following equations represent hyperbolas:

$$yx = f^2;\ \frac{ly^2}{g} = x^2 - f^2;\ y^2 - f^2 = \frac{lx^2}{g};$$

and that $\frac{ly^2}{g} = f^2 - x^2$

represents an ellipse. Then it is shown, by means of some examples, how to reduce a given equation to one of these standard forms (if possible), including a detailed description of the curve under consideration.

Here also the concept of "focus" (umbilicus) of a hyperbola or an ellipse is introduced in the usual way by characterizing a hyperbola (ellipse) as the locus of the points with the property that the difference (sum) of the distances from this point to the foci has a constant value.

Chapter IV of *Liber Secundus* is the only one with a heading: *Regula Universalis Inveniendi ac Determinandi Loca Quaelibet Plana et Solida.* [*General Rule to Find and to Determine Arbitrary Plane and Solid loci*].

By "plane and solid loci" he means curves of the first or second degree. This chapter contains a nearly exhaustive classification of these curves and it is really very complicated, mainly because–as we said before–only positive coefficients are taken into consideration and because there is no y-axis.

For further details about *Liber Secundus,* the reader is referred to my annotated translation of this book, which is in preparation.

vi. The work of Jan de Witt has often been compared with John Wallis's (1616–1703) *Tractatus de Sectionibus Conicis* [*Treatise on Conic Sections*], which appeared in 1655. In this work the conics, for the first time in history, are formally *defined* by means of algebraic equations, without any appeal to their geometric meaning. Wallis's definitions are:

$$p^2 = ld \text{ (parabola)};\ e^2 = ld - \frac{ld^2}{t} \text{ (ellipse)};$$

$$h^2 = ld + \frac{ld^2}{t} \text{ (hyperbola)}.$$

In all these cases, the abscissa (x-coordinate) is denoted by d (for distance) and the ordinate (y-coordinate) is denoted by p (parabola), e (ellipse) or h (hyperbola). The letter l represents the latus rectum and t the latus transversum (see also Appendices A and B).

Of course, these defining equations did not appear out of the blue: they were suggested by the characteristic properties of the conics (συμπτώματα) that had been derived by Apollonius (see Appendix B).

Wallis showed that the curves defined by these equations are indeed the well-known parabola, ellipse, and hyperbola; then he derived a number of properties from these equations in a purely algebraic way. In his work, however, there is not a trace of a classification; this had already been done by de Witt.

As stated earlier, Jan de Witt completed his work in about 1649; as Wallis's work did not appear until 1655, we have no reason to assume that de Witt could have been influenced by John Wallis, particularly since their points of departure, in particular their definitions of a conic, are different.

9. The present translation of *Elementa Curvarum Linearum–Liber Primus* is based on the printed text of the second edition, added to the third edition of van Schooten's *Geometria* (1683). In fact, this is the text of the first edition (1659), corrected in accordance with the errata that had been added to the first edition on a separate page. In section 6, we mentioned the letters exchanged between van Schooten and de Witt concerning these errata.

During the process of translation, the translator spotted some more misprints. Here they are; the page numbering is that of the original Latin text:

page [183], line 1 from the top: for C, read c.
page [197], line 3 from the top: for ac, read ad.
page [201], line 7 from the bottom: for AH, read AB.
page [201], line 16 from the bottom: for 2, read 12 (in superscript).
page [209], line 3 from the top: for aequidistant, read aequidistantia.
page [209], line 93 from the top: for ttansversa, read transversa.
page [213], line 2 from the top: for ad, read ac.
page [214], line 2 from the top: for sic, read sit.
page [225], line 3 from the bottom: for ocrurrant, read occurrant.
page [231], line 12 from the bottom: for ad, read at.
page [240], line 5 from the bottom: for intersectionem, read intersectionum.

10. Finally, some remarks on the translation. I have intended the translation to adhere to the Latin text as closely as possible. This means, first of all, that the theorems and their proofs, which are in given in *Liber Primus* purely verbally, without using any symbolic language or formula, have been rendered in the same way.

In the notes, however, I rewrite many proofs in the conventions of our modern mathematical language. In the translation, these notes are numbered in the following way: [2.14], which means note 14 to chapter II.

Jan de Witt inserted into his text three kinds of marginal notes:

a. References to the *Elementa* by Euclid, by means of superscript numbers. Only the first reference (p. [164], 1) mentions the source: "per Cor. 8 sexti Eucl," i.e., "because of Corollary 8 of the sixth book of Euclid." The other ones have a standard form. A typical example: "per

9 quinti" refers to the ninth chapter of the fifth book of the *Elementa* by Euclid. In the translation, this reference is briefly formulated as "V, 9," etc. It can be remarked that Jan de Witt justifies his steps meticulously. Thus, the above marginal note refers to the theorem that $a : b = c : b$ implies: $a = c$ (!).

b. References to the work *Elementa Curvarum Linearum* itself. An example of this: "per 2 Cor. 13 et 3 Cor. 14 hujus" refers to Corollary 2 of Proposition 13 and Corollary 3 of Proposition 14 "of this work." For the comfort of the reader, this last, somewhat vague, specification has been replaced by a precise identification of the page. This reference is represented as : "Prop. 13, Cor. 2, and Prop. 14, Cor. 3, p. [221]."
c. References by means of tiny superscript letters. These generally contain a technical clarification of the text, often discerning the various possibilities. These have been translated as accurately as possible and offer good services at reading this work.

The reader finds the notes at the bottom of the pages.

A typical feature of de Witt's method is that he seldom explicitly gives the place of the figure under discussion. This gap could have been filled in the translation, but has not been, partly in order to prevent the text from getting too crowded with references, but primarily to preserve the flavour of the original text.

Another feature: Jan de Witt never refers to a straight line or a curve by means of one letter, say k, l, or m, but always uses at least two letters, usually capitals; these same letters are used afterwards to denote specific points. An example: page [193] introduces a hyperbola, whose branches are called IC and HE; here the points I, C, H, and E are still unspecified. Next an arbitrary transverse diameter is drawn, and then the letters C and E are used to denote the vertices of this diameter, so that it is only here that the points C and E receive their exact definition.

As stated earlier, the translation keeps as closely to the Latin text as possible. Nevertheless it turned out to be necessary to split up long sentences into smaller parts.

A typical example is given by a sentence on page [228], which occupies twenty-one lines! In such cases, subordinate clauses are rendered as main sentences (mutatis mutandis).

Another construction, which occurs very often, is the following (e.g., page [194]):

> *Ductis (...) asymptotis (...) ductaque a C (....) recta CB (...) sumatur inter AB, BC media proportionalis AD.*

Frans van Schooten Jr. (1615–1660).

Hiëronymus van der My (1687–1761) pinxit.
Modeled to a portrait by Adriaen Cornelisz Beeldemaker (1618–1709).
Source: Collection Academisch Historisch Museum, Leiden

Here the translation reads:

> *First we draw the asymptotes and from C we draw the line segment CB, then we take the mean proportional AD between AB and BC.*

Nevertheless, the translation does preserve, at times, a solemn and somewhat archaic style.

Those who want to get a quick impression of the contents of the work are referred to the summary.

2

Summary

In this summary the theorems and their corollaries are given without proofs, restated in our modern notation. In *Liber Primus* Jan de Witt operates purely verbally, without using our common symbols $+, -, \times, :, =$, etc.

The essence of the statements, however, has been preserved. For the proofs we refer to the text, the translation, the notes, and the appendices. This summary aims only at giving a global survey of the contents of the book.

Chapter I

Jan de Witt starts with a fixed straight line r (the directrix), a fixed point T (the pole), and a fixed straight line m, passing through T and intersecting the directrix at a point A; TA is called the interval, the parameter, or the latus rectum. The curve concerned is generated in the following way: an angle α (the moving angle), with vertex T and arms s (the passive arm) and w (the active arm), rotates about the pole T. During this process, s intersects the directrix at a variable point P. A straight line (the generator b) is drawn through P, parallel to TA. The curve we are interested in is described by the moving point of intersection S of the generator b and the active arm w of the rotating angle α (see Figure 2.1).

In chapter I we consider only the case in which the angle TAP ($= \beta$) is equal to the angle α, and our statements are valid in this situation only. In this case the point S turns out to describe a parabola (see Figure 2.2).

The case in which the angle TAP differs from the angle α is postponed until chapter IV; in that case the point S describes a hyperbola. Very

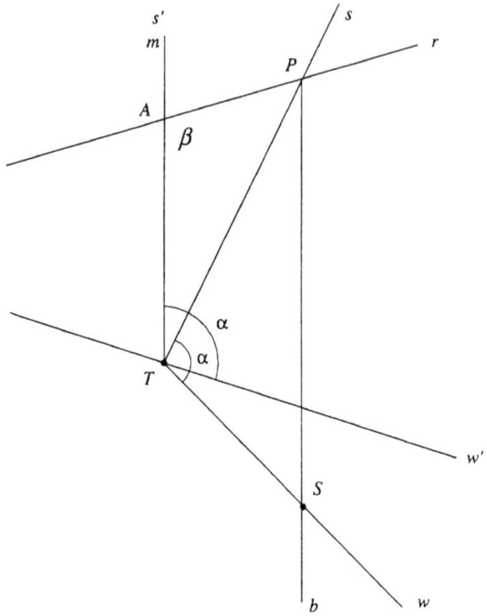

FIGURE 2.1

important in this context is the concept: "ordinate-wise applied." We say that a straight line d is ordinate-wise applied to m if d is parallel to the active arm in the initial position (w'), i.e., in the position in which the passive arm of α coincides with m. It will turn out that every chord that is ordinate-wise applied to m is bisected by m. This is why we call m a diameter. As a consequence, the active arm in the initial position is a tangent to the parabola at T. The essence of chapter I consists of two theorems:

i. The first theorem states that the curve described in this way has the characteristic property of a parabola (roughly speaking, $y^2 = px$).
ii. The second theorem states that every straight line parallel to m is also a diameter, since it bisects all chords parallel to the tangent at the corresponding vertex, i.e., the point of intersection of this diameter and the parabola.

N.B. This diameter has its own interval, its own moving angle, and its own directrix.

Theorem I. If one selects on the curve a point S, and passes through S a straight line ordinate-wise applied to m and intersecting m at the point K, then the following holds (see Figure 2.2):

$$SK^2 = TA.KT$$

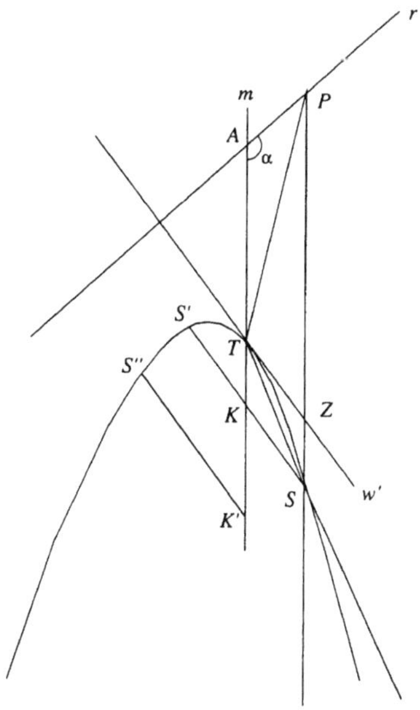

FIGURE 2.2

For this reason we call TA the parameter or the latus rectum. De Witt associates "his" parabola with the parabola defined by Apollonius (see Appendix B). In passing he remarks that the diameter is an axis if the moving angle is a right angle.

COROLLARY 1. A straight line parallel to a diameter intersects the curve at only one point.

COROLLARY 2. If one draws a straight line from the pole T to a point S on the curve, then TS lies wholly inside the curve and the produced part of TS lies wholly outside the curve.

COROLLARY 3. When the point S traverses the curve while the moving angle rotates about T, then the angle between TS and m will become smaller than any prescribed angle, but TS will never coincide with m.

COROLLARY 4. Every straight line intersecting the diameter m will, being produced, also intersect the curve. Here, as usual, the diameter is that part of m that lies inside the curve.

COROLLARY 5. Every chord that has been ordinate-wise applied to m is bisected by m. This explains the name diameter.

COROLLARY 6. Conversely, m bisects only those chords that are ordinate-wise applied to m.

COROLLARY 7. The ordinate-wise applied segments SK and $S''K'$ (see Figure 2.1) satisfy the relation:

$$SK^2 : S''K'^2 = TK : TK'.$$

Corollary 8. The active line in the initial position touches the curve only at T. Moreover, it has no other point in common with the curve.

Corollary 9. Apart from the active line in the initial position, no other straight line touches the curve at T.

Corollary 10. A method to construct the curve pointwise, when given the position of the straight line m, its vertex, the corresponding interval (latus rectum, parameter) TA, and the angle between m and the straight lines that are ordinate-wise applied to m.

Theorem II. A straight line passing through an arbitrary point (e.g., T') on the curve, and parallel to the diameter that was used in defining the curve (in this case m), is a diameter as well, with T' as its vertex (see Figure 2.4).

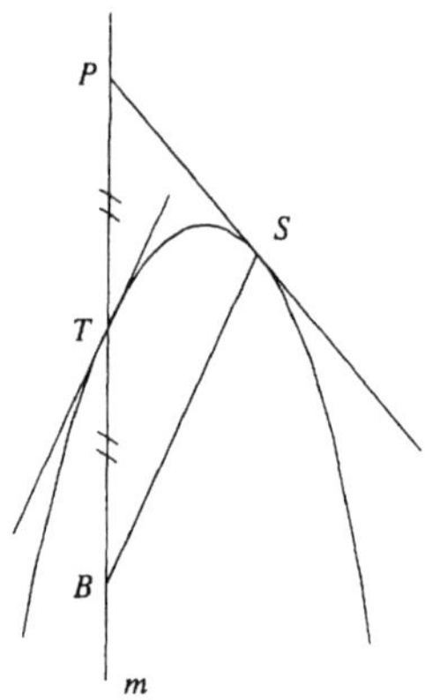

Figure 2.3

Corollary 1. A straight line connecting the midpoints of two mutually parallel chords is a diameter.

Corollary 2. A tangent and a straight line passing through the point of contact and ordinate-wise applied to the diameter, intersect the diameter at points that have the same distance to the vertex. In Figure 2.3 the diameter is represented by m; the point of contact is S; SP and SB are the tangent and the applied line, respectively. According to this corollary PT is equal to TB.

Corollary 3. The construction of a tangent at a point on the parabola and from a point outside the parabola.

Corollary 4. This corollary gives the magnitude of the parameter p' corresponding to the new diameter defined in Theorem II. In Figure 2.4

the original diameter is represented by m, its vertex is T; m' is the new diameter (parallel to m), T' is its vertex. The straight line $T'R$ touches the curve at T' and intersects the diameter m at the point R. The new parameter p' satisfies

$$TR : T'R = T'R : p'.$$

COROLLARY 5. In this corollary the following problem is solved: We are given the position of a diameter of a parabola, its vertex, the corresponding parameter, and the angle between this diameter and the straight lines that are ordinate-wise applied to it. It is required to construct a diameter such that the straight lines that are ordinate-wise applied to it make a given angle with this diameter, and to find the corresponding parameter.

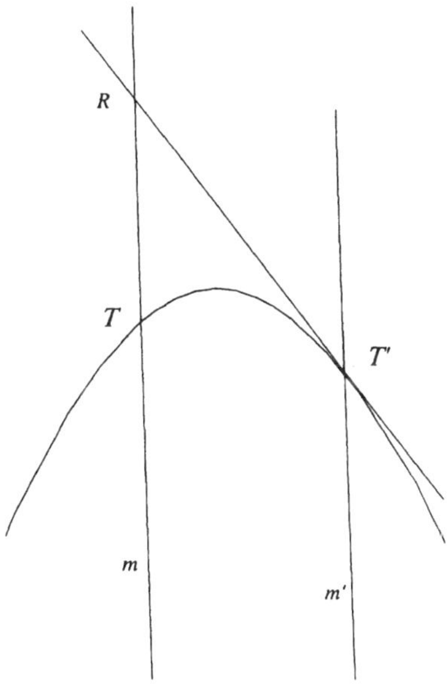

FIGURE 2.4

Chapter II

The curve that is studied in this chapter is not generated by a rotating angle and a moving straight line, but by a rotating straight line and a moving angle (see Figure 2.5). De Witt starts again with a fixed straight line r (the directrix), a fixed point T (the pole) and a straight line b (the generator) passing through the pole T. This generator rotates about the pole and always passes through the endpoint A of one arm BA (of constant length) of a given angle. The generator moves this arm (the passive arm)

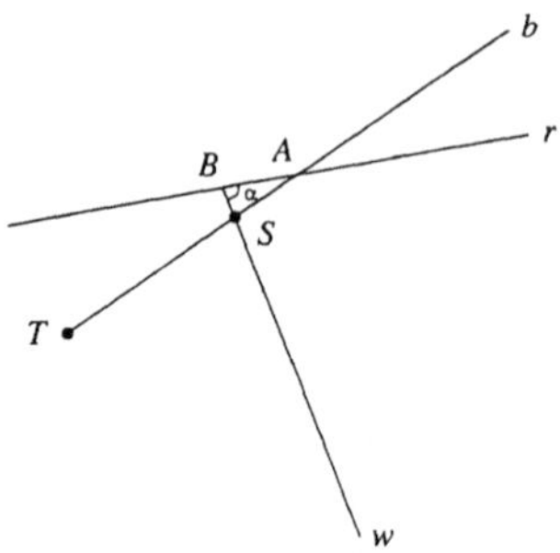

FIGURE 2.5

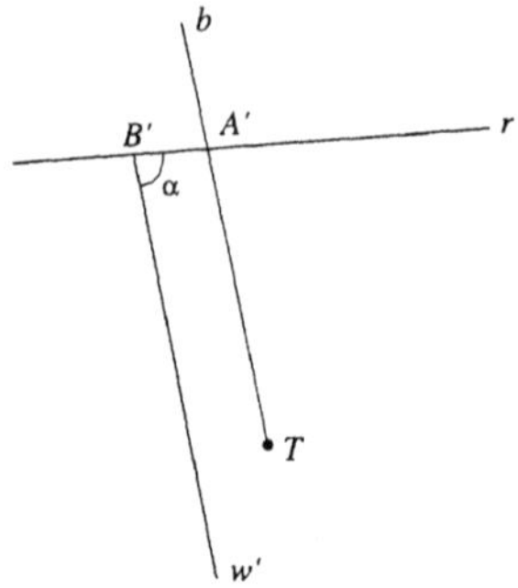

FIGURE 2.6

along the directrix during its rotation. De Witt is interested in the curve described by the point of intersection S of the rotating generator b and the other arm w (the active arm) of the moving angle. It will turn out that the curve generated in this way is a hyperbola. When the active arm w

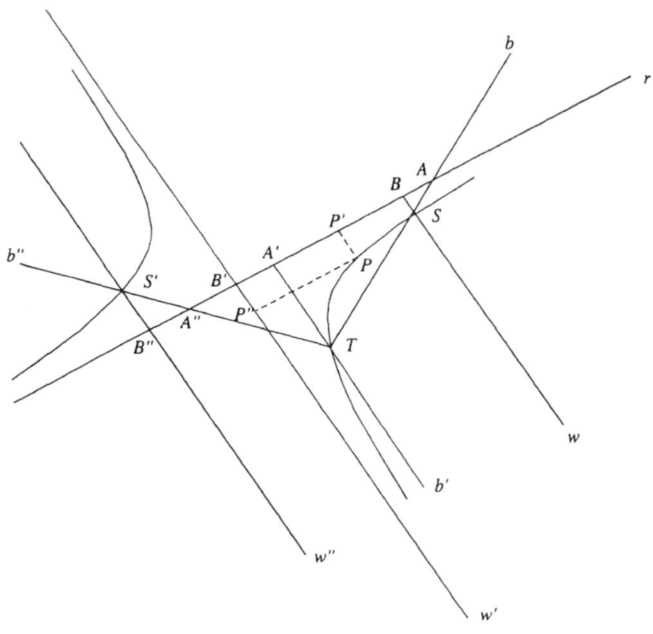

FIGURE 2.7

is parallel to the generator b, then we say that both straight lines are in the initial position (see Figure 2.6). Next, we define two distinct intervals: first, the constant length AB of the arm that is moved along the directrix and second, the distance TA' from T to the directrix, measured along the generator in the initial position. It turns out that the curve described in this way is a hyperbola. Its two branches are called the opposite hyperbolas (see Figure 2.7). De Witt uses the term hyperbola, even when he refers to one branch.

Chapter II consists of the following theorems and corollaries (the numbering is the same as in the Latin text).

Theorem III. If one selects an arbitrary point P on the curve and passes through P the straight lines PP' and PP'' parallel to the active line in the initial position and to the directrix, respectively (see Figure 2.7), then

$$PP'.PP'' = TA'.A'B'$$

or, putting $$PP'' = x, \ PP' = y, \ TA' = a, \text{ and } A'B' = AB = b,$$

$$xy = ab.$$

So the curve turns out to be a hyperbola, with the directrix r and the active line in the initial position as its asymptotes. The product ab is called the power of the hyperbola (Lat. *potentia*). The point of intersection of the asymptotes is called the center.

COROLLARY 1. The distance between a suitably selected point on the hyperbola and an asymptote can be made arbitrarily small.

COROLLARY 2. We select a branch of a hyperbola and a point inside the angle opposite to the angle in which this branch is situated. If we draw a straight line through this point that either passes through the center or intersects one of the asymptotes, then this line will intersect the hyperbola at just one point.

COROLLARY 3. The active arm intersects the hyperbola and every straight line parallel to the asymptote that is not parallel to the active arm at just one point.

Theorem IV. A straight line that passes through two points on a hyperbola, or is tangent to it, intersects both asymptotes inside the angle containing this hyperbola.

Theorem V. If one selects two points on one branch of a hyperbola, or on both branches, and passes a straight line, or two parallels, through them, then the rectangles contained by the segments, intercepted on both sides between the curve and an asymptote, will be equal.

NOTE. An explanation by a figure is required. In Figure 2.8, $SAT \parallel S'A'T'$

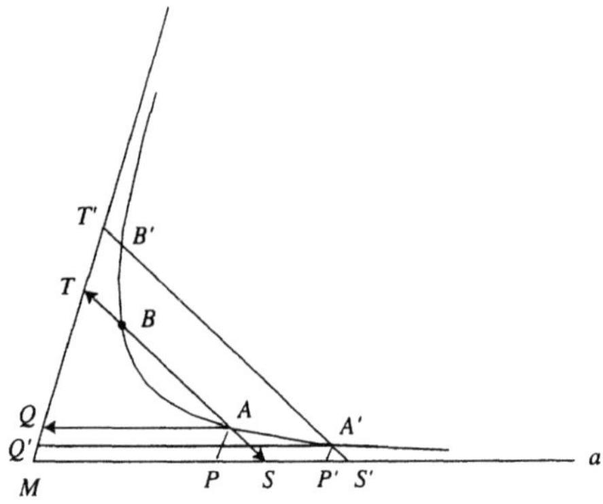

FIGURE 2.8

and $AS.AT = BT.BS$, but also $AS.AT = A'S'.A'T'$. If the points concerned lie on different branches of the same hyperbola, then analogous theorems hold.

COROLLARY 1. Suppose one passes a straight line through the center of a hyperbola intersecting both branches at P and Q (see Figure 2.9). If one then draws a straight line parallel to it, intersecting both branches at A, resp. B, but intersecting the asymptotes at S and T, then

$$PM^2 = AS.AT = BT.BS = QM^2.$$

In this case $PM = QM$; this justifies the name "the center."

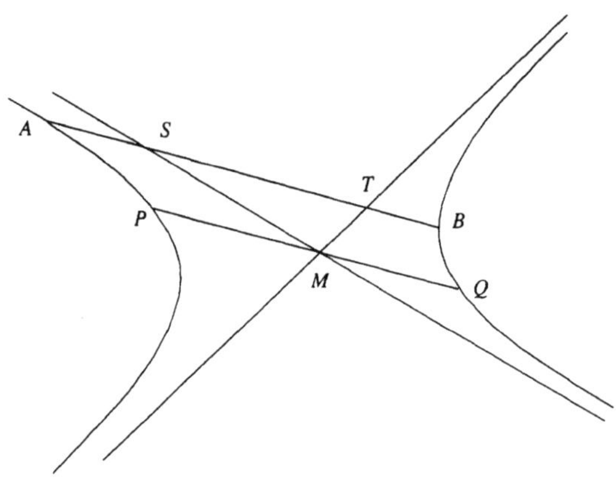

FIGURE 2.9

Corollary 2. (See Figure 2.9.) As a consequence $AS = TB$ and $AT = BS$.

Corollary 3. A straight line connecting two points on one hyperbola or on two opposite hyperbolas has no other point in common with the hyperbola.

Corollary 4. The converse of Theorem V is also true (see Figure 2.8): if under the same circumstances it is given that $AS.AT = A'S'.A'T'$, and if one of the points A and A' lies on the hyperbola, then the other point lies on the hyperbola as well.

Corollary 5. A straight line passing through the center of a hyperbola and bisecting a chord will also bisect all chords parallel to that chord.

At this point Jan de Witt introduces some technical terms.

a. A line passing through the center is called a diameter, in particular:
   - an intercepted (transverse) diameter, or simply a diameter, is a diameter that intersects the hyperbola.
   - a second diameter is a diameter that does not intersect the hyperbola.
b. Parallel chords that are bisected by a diameter are said to be ordinate-wise applied to this diameter. If these chords are perpendicular to this diameter, then this diameter is called an axis.
c. If, however, a second diameter is parallel to the chords that are ordinate-wise applied to an intercepted diameter, then it is said that these two diameters are mutually conjugate.

Note. From theorem VI it is clear that the relation "conjugate to" is a symmetric relation.

Corollary 6. A straight line passing through the center and bisecting a chord will bisect only those chords that are parallel to this one.

Corollary 7. A straight line passing through the midpoints of two parallel chords inside one branch of a hyperbola, or lying between two branches, will pass through the center and therefore be a diameter. From this it is clear how to find arbitrary diameters of a given hyperbola or of two branches, and how to construct the corresponding ordinate-wise segments and the center.

Theorem VI. A line segment passing through a point on a hyperbola, bounded by the asymptotes, and bisected at this point is tangent to the hyperbola at this point. Conversely, a line segment that is bounded by the asymptotes touching the hyperbola at a point is bisected at this point.

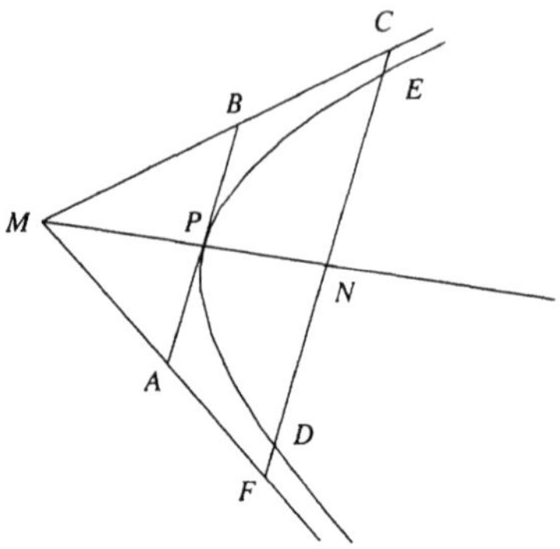

FIGURE 2.10

COROLLARY 1. (See Figure 2.10.) Here APB touches the hyperbola at P; the segment $FDEC$ is parallel to APB. The points A, B, C, and F lie on the asymptotes; D and E lie on the hyperbola. Then the following holds:

$$DC.DF = EF.FD = EF.EC = EC.CD = AP^2 = BP^2.$$

COROLLARY 2. A line segment bounded by the asymptotes, passing through the endpoint of a diameter, and parallel to a chord bisected by this diameter touches the hyperbola at the endpoint of this diameter.

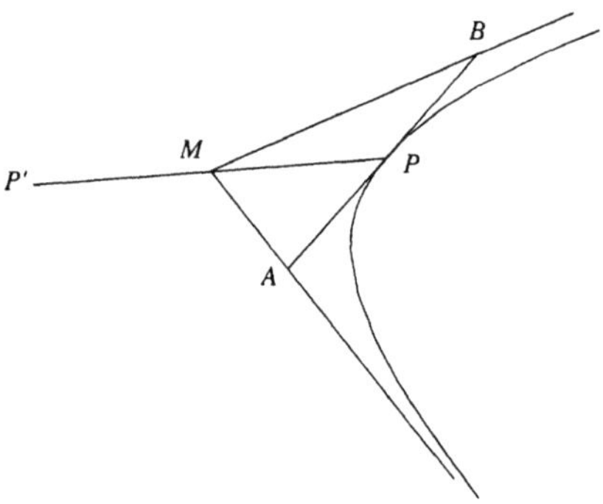

FIGURE 2.11

COROLLARY 3.

a. All chords parallel to a tangent are bisected by the diameter that passes through the point of contact.

b. There exists only one tangent at a point on a hyperbola.

At the end of this corollary Jan de Witt assigns a length to an intercepted (transverse) diameter and to its conjugate (second) diameter (see Figure 2.11); he also introduces the parameter.

a. Let MP be a transverse diameter, P its vertex, and P' the point of intersection with the opposite hyperbola; then PP' $(= 2MP)$ is the length of the transverse diameter.
b. The diameter conjugate to PP' is parallel to the tangent at P. The length of this conjugate diameter is defined as the length of that part of the tangent at P that is intercepted between the asymptotes. So in Figure 2.11 this length is the length of AB.
c. The parameter (latus rectum) p corresponding to the transverse diameter PP' and its conjugate AB (in this order) is defined as the third proportional to PP' and AB, i.e., by the proportion $PP' : AB = AB : p$.

Theorem VII.

a. A straight line passing through the endpoint of a transverse diameter and parallel to the tangent at its vertex is tangent to the opposite hyperbola.
b. If one selects an arbitrary transverse diameter and applies a straight line ordinate-wise to its conjugate diameter, then this straight line is parallel to the selected transverse diameter. See Figure 2.12. Here PP' is the selected transverse diameter and QQ' its conjugate diameter; AB has been ordinate-wise applied to QQ'. Then $AB \parallel PP'$.

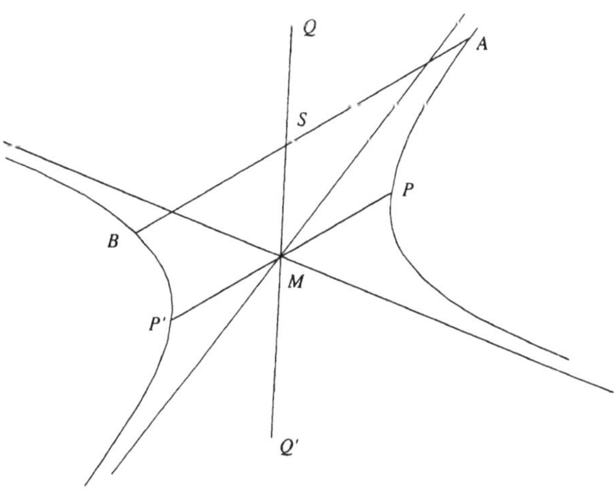

FIGURE 2.12

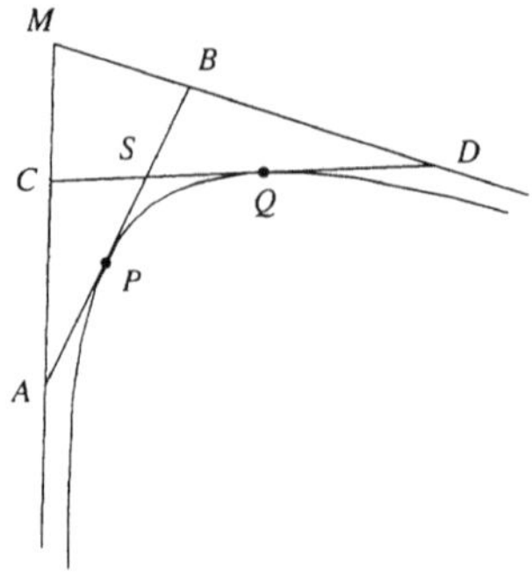

FIGURE 2.13

Problem 1. It is required to construct the axes of a hyperbola, when the length and the position of two mutually conjugate diameters are given.

Theorem VIII. (See Figure 2.13.) In this figure MA and MD are the asymptotes of a hyperbola. APB and CQD are tangent to the hyperbola at P and Q, respectively. Then the following hold:

a. Area ΔMAB = area ΔMDC
b. $MB.MA = MC.MD$
c. $AC : CM = DB : BM$.
d. $AS : SB = DS : SC$
e. $AP : PS = DQ : QS$

Theorem IX. (See Figure 2.14.) In this figure $P'MP$ is a transverse diameter of length $2a$. APB has been ordinate-wise applied to $P'MP$ and therefore touches the hyperbola at its vertex P. The length of APB is $2b$. CDE has been ordinate-wise applied to $P'MP$. The length of PH is equal to the corresponding parameter $(2b^2/a)$. Then the following holds:

$$AB^2 : P'P^2 = p : 2a = ED^2 : P'D.PD.$$

COROLLARY 1. A method to construct the asymptotes of a given hyperbola. Note: "Given" means that the curve has been drawn in the plane.

COROLLARY 2. In Figure 2.14 we have: $ED^2 = PD.DK$.

NOTE. At this point Jan de Witt discusses the relationship between "his" hyperbola and Apollonius's hyperbola and shows that they are identical. See also note [2.29] and Appendix B.

COROLLARY 3. (See Figure 2.14.) Here ST is parallel to ED, so

$$ST^2 : P'T.PT = ED^2 : P'D.PD.$$

Theorem X. (See Figure 2.15.) In this figure MB and MC are the asymptotes of a hyperbola AP, MP is an arbitrary transverse diameter, and P is its vertex. BAC is the tangent to the curve at A; AD has been

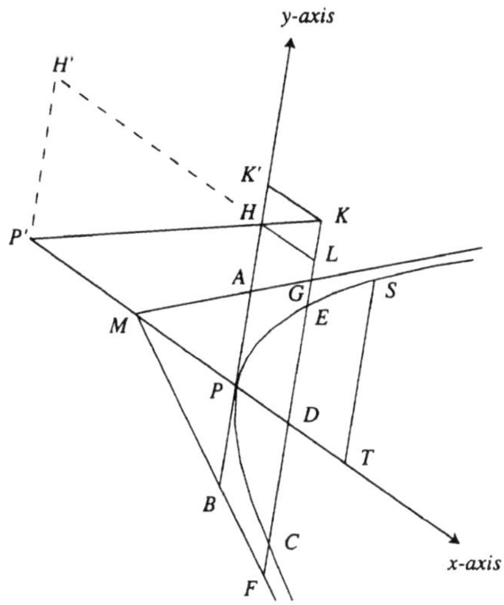

FIGURE 2.14

ordinate-wise applied to the diameter MP; BAC and MPD meet each other at Q. Then $MP^2 = MQ.MD$.

Theorem XI. (See Figure 2.16.) In this figure MB and MC are the asymptotes of a hyperbola AP. MP is an arbitrary transverse diameter and P is its vertex. BAC is the tangent to the curve at A. AD has been ordinate-wise applied to the second diameter MT, which is conjugate to MP; BAC and MT meet each other at Q. Then the following holds:

$$PS^2 = MQ.MD.$$

From theorem X and theorem XI a method is deduced to construct the tangent(s) to this curve from a point (on or outside a given hyperbola).

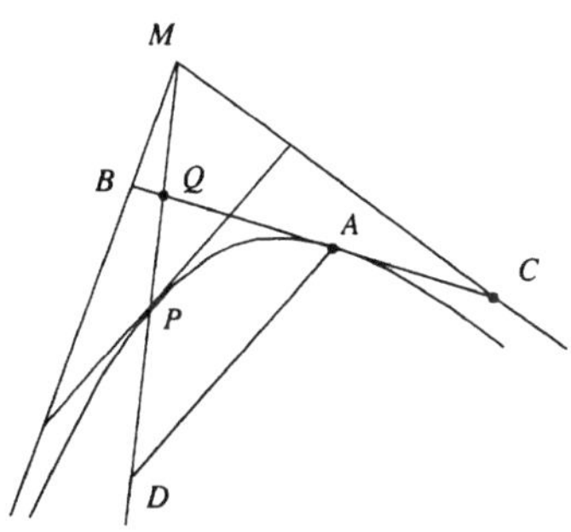

FIGURE 2.15

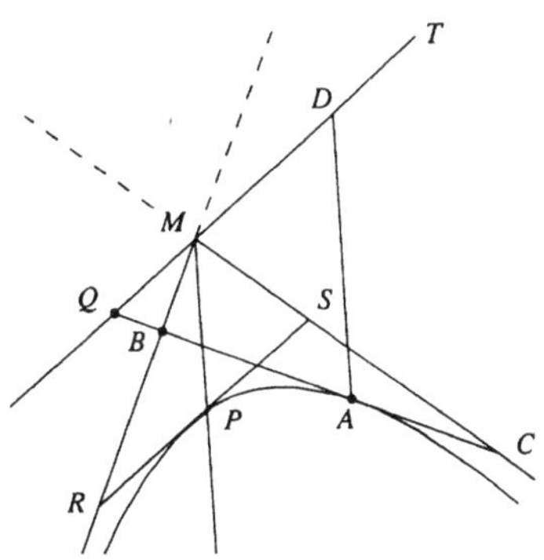

FIGURE 2.16

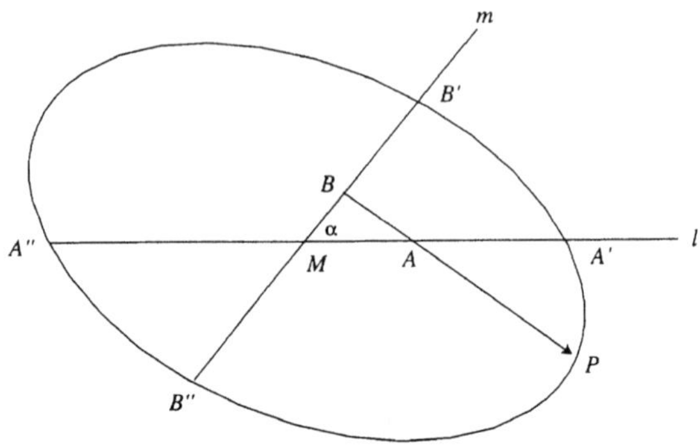

FIGURE 2.17

Chapter III

In this chapter a curve is described that turns out to be an ellipse. De Witt starts with an angle with M as its vertex and l and m as its arms (see Figure 2.17). A line segment AB slides with the endpoints A on l and B on m. On AB or on the produced part of AB (or BA) a fixed point P is selected. De Witt is interested in the locus of P when A and B traverse the whole l and m respectively. The point P is called the active point, or simply the point; the segment AB is called the generator, the segments PA and PB are called the intervals, and M is called the center. If one

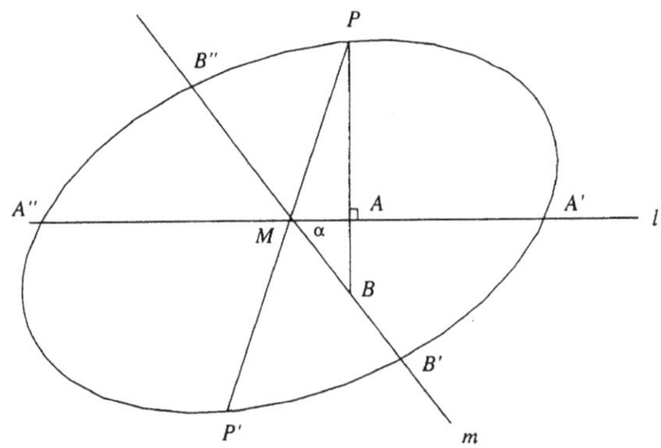

FIGURE 2.18

measures the segment PB (with endpoints $A'$and A'') along the arm l on both sides from M, then the segment $A'A''$ is called the directrix (see Figure 2.18). In the same way we can measure the segment PA (with endpoints B' and B'') along m, on both sides of M . Then this segment is also called the (corresponding) directrix, but we have to make a choice. We say that the generator AB is in the initial position when it is perpendicular to the

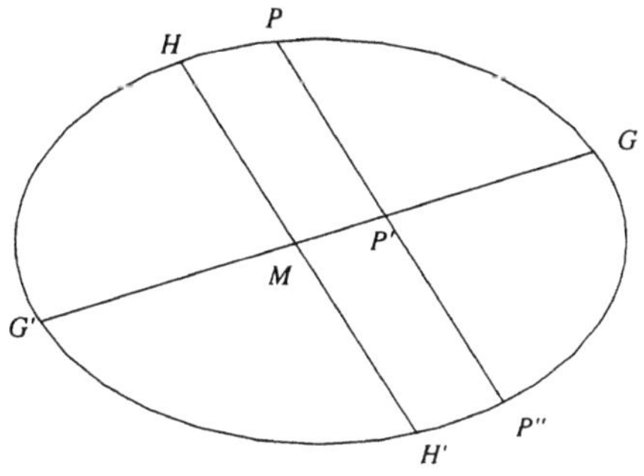

FIGURE 2.19

directrix. Twice the length of MP in this position is called the secant. When AB is in this position, it is also said that the active point P is in the initial position (see Figure 2.18).

Theorem XII. If one draws a straight line parallel to the secant through an arbitrary point P on the curve and intersecting the directrix at P', then the following holds (see Figure 2.19):

$$PP'^2 : G'P'.P'G = MH^2 : MG^2.$$

NOTE. This theorem induces Jan de Witt to compare "his" ellipse with Apollonius's ellipse, and he shows that they are identical. For details see notes [3.5], [3.10] and Appendix B. At this point Jan de Witt introduces some technical terms.

a. The directrix and the secant are called conjugate diameters.
b. Any line segment passing through the center and bounded by the ellipse on both sides is called a diameter.
c. Two arbitrary diameters HH' and GG' (see Figure 2.20) are mutually conjugate if, for the segment PP' (P on the curve and P' on the directrix, $PP' \parallel HH'$), the following holds:

$$PP'^2 : G'P'.P'G = MH^2 : MG^2.$$

In this case $G'G$ is called a transverse diameter and $H'H$ is called the corresponding second diameter. If two conjugate diameters are mutually perpendicular, they are called the *axes* of the ellipse.
d. The parameter p corresponding to a transverse diameter GG' and its second diameter HH' (in this order), is defined as the third proportional to GG' and HH', which means that, if $GG' = 2a$ and $HH' = 2b$, the following holds:

$$2a : 2b = 2b : p, \text{ so } p = 2b^2/a.$$

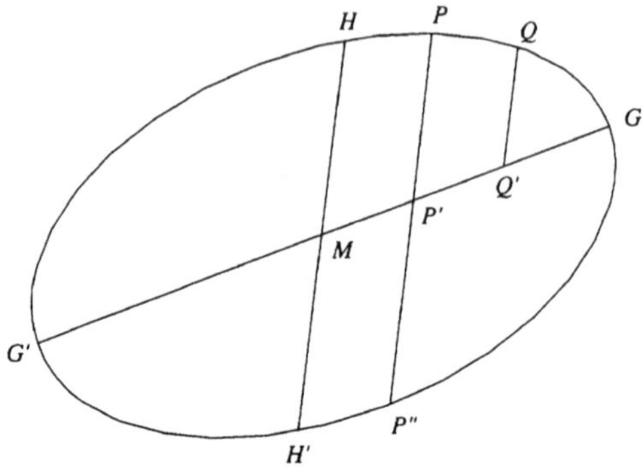

FIGURE 2.20

COROLLARY 1. The transverse axis of two conjugate axes is also a second axis, and conversely.

COROLLARY 2. The straight line passing through the active point in the initial position and parallel to the directrix touches the ellipse at this point and has no other point in common with the ellipse.

COROLLARY 3. If one applies line segments from the points P' and Q' (on a diameter of an ellipse) to the points P and Q on the circumference, parallel to its conjugate diameter, e.g., $P'P$ and $Q'Q$ as in Figure 2.20, then

$$P'P^2 : Q'Q^2 = G'P'.P'G : G'Q'.Q'G.$$

COROLLARY 4. Chords that are ordinate-wise applied to an axis or to a diameter are bisected by this diameter.

COROLLARY 5. Chords that are ordinate-wise applied do not intersect the ellipse at more than two points. Jan de Witt tacitly assumes that a straight line is ordinate-wise applied to a diameter if it is parallel to the corresponding second diameter.

COROLLARY 6. (See Figure 2.21.) In this figure OB and CD are mutually perpendicular diameters (axes). BF touches the ellipse at B and its length is equal to the parameter corresponding to the diameters OB (transverse diameter) and CD (second diameter), so $CD^2 = OB.BF$. The point P is an arbitrary point on the ellipse and PS is perpendicular to OB. Then

$$PS^2 = QS \cdot SB.$$

NOTE. Jan de Witt assumes without proof that this theorem holds for arbitrary conjugate diameters. Moreover, he refers at this point to the main property of an ellipse, as defined by the "Ancient Geometers."

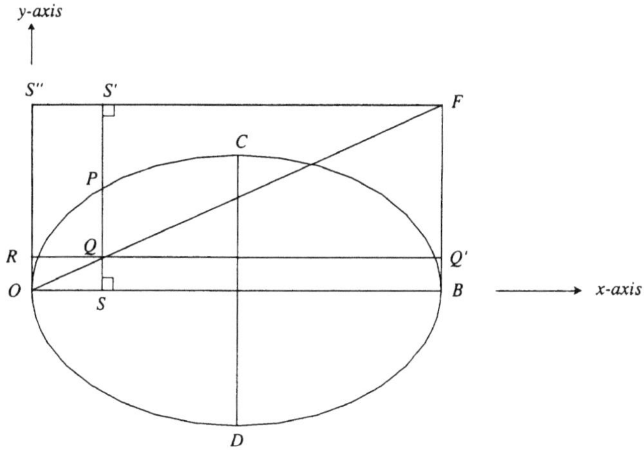

FIGURE 2.21

COROLLARY 7. A method to describe an ellipse, when given a pair of conjugate diameters.

NOTE. In fact, Jan de Witt constructs from these data the angle and the intervals.

Theorem XIII. In an ellipse every diameter is a transverse diameter and has a corresponding second diameter.

COROLLARY 1. A method to construct the axes of an ellipse when two conjugate diameters are given.

COROLLARY 2. In an ellipse every diameter is bisected at the center.

COROLLARY 3. In an ellipse, the transverse diameter of a pair of conjugate diameters is also a second diameter, and conversely.

COROLLARY 4. A straight line passing through the endpoint of a transverse diameter parallel to the corresponding second diameter is the tangent to the ellipse at this point and has no other point in common with the ellipse.

Theorem XIV. A chord connecting two points on an ellipse and bisected by a diameter either passes through the center or is ordinate-wise applied to this diameter.

COROLLARY 1. A diameter in an ellipse that bisects a chord (not passing through the center) will also bisect all chords parallel to that chord.

COROLLARY 2. A straight line in an ellipse that bisects two parallel chords passes through the center and is therefore a diameter.

COROLLARY 3. A chord connecting two arbitrary points on an ellipse lies entirely inside the ellipse.

Problem II. In an arbitrarily given ellipse, construct the conjugate diameter corresponding to a given diameter.

COROLLARY. The construction of the tangent to an ellipse at a given point on the curve.

Theorem XV. The straight line passing through a point on an ellipse and ordinate-wise applied to the diameter through this point is the only tangent at this point.

COROLLARY. In an ellipse all chords parallel to a tangent are also parallel to the diameter conjugate to the diameter through the point of contact. They are therefore ordinate-wise applied to the diameter through the point of contact and are bisected by it. Conversely, a straight line passing through the endpoint of a diameter and parallel to a chord that is bisected by this diameter touches the ellipse at this vertex.

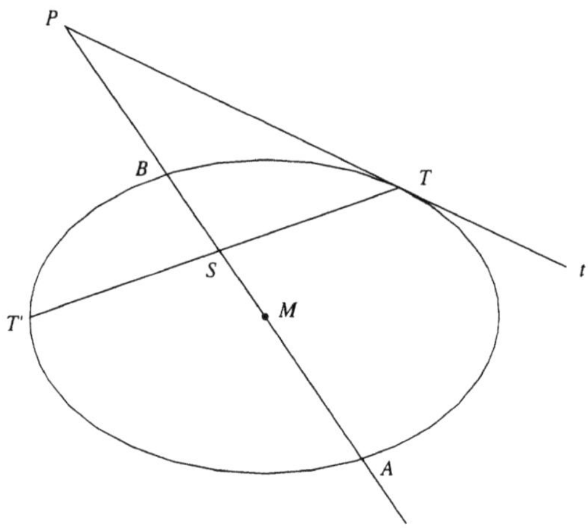

FIGURE 2.22

Theorem XVI. (See Figure 2.22.) Here AMB is an arbitrary diameter of an ellipse with center M. The tangent at T intersects the produced part of AMB at P. The chord TT' is ordinate-wise applied to this diameter, intersecting AMB at S. Then $MP.MS = MB^2$. Conversely: If TS has been ordinate-wise applied to the diameter AMB and if $MP.MS = MB^2$, then PT touches the ellipse at T.

COROLLARY. A method to construct a tangent to an ellipse from a point on this curve or outside this curve.
NOTE. Strangely enough, de Witt restricts himself to the construction of only one tangent.

Chapter IV

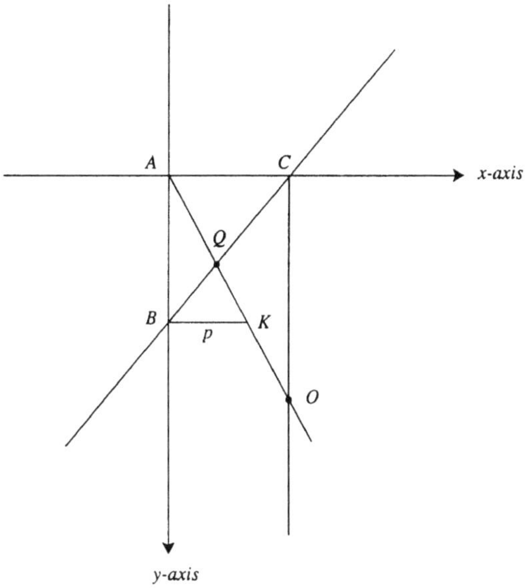

FIGURE 2.23

This chapter is the first chapter with a heading:

Another way to describe a parabola, a hyperbola or an ellipse in the plane.

For obvious reasons, it is made up of three parts.

A. In the first part, a new method is explained to draw a parabola in the plane. In modern notation its essence is the following (see Figure 2.23): the point A is the origin of an orthogonal coordinate system. ABC is a triangle whose (mutually equal) sides AC and AB lie on the positive x-axis and the positive y-axis, respectively. The hypotenuse BC always moves parallel to itself and drags along a straight line passing through C and parallel to the y-axis. The point B drags along a line segment BK of constant length p, always parallel to the x-axis. A straight line rotates about the origin A and always passes through the moving endpoint of BK. De Witt is interested in the locus of the point of intersection O of the rotating line through A and the line that is dragged along parallel to the y-axis by the moving point C. The curve described in this way turns out to be a parabola. Incidentally de Witt remarks that the point of intersection Q of BC and AK describes a hyperbola.

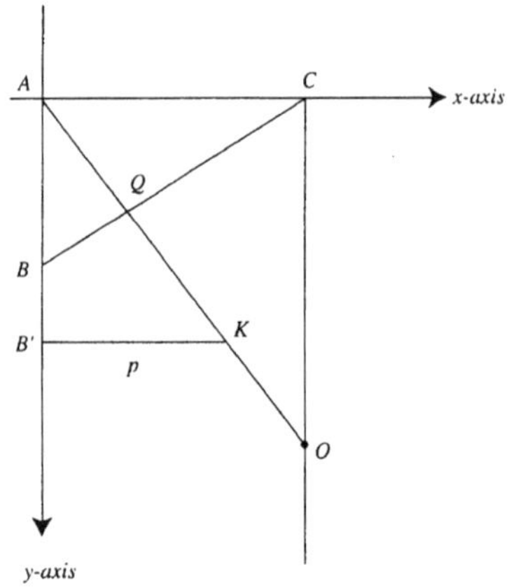

FIGURE 2.24

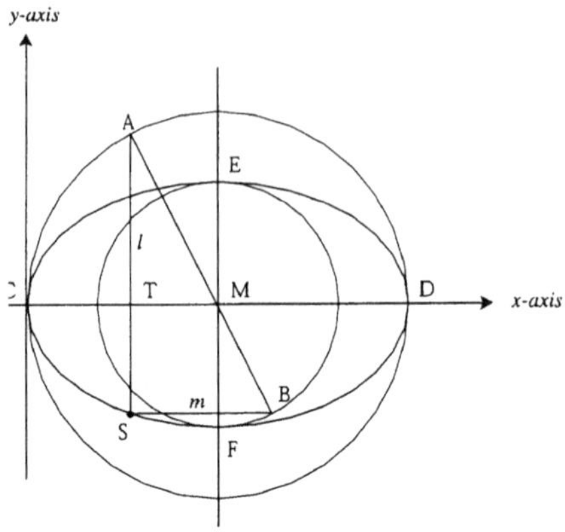

FIGURE 2.25

Moreover, he says that similar conclusions can be drawn if the triangle ABC is not isosceles, and also if the segment of length p starts not at the point B but at the point B' on the y-axis at a fixed distance from B (see Figure 2.24).

B. Here Jan de Witt goes back to the beginning of chapter I. There a curve is described by means of a given directrix, interval and moving angle, according to a fixed rule (see Figure 2.1). In chapter I de Witt restricts himself to the case in which the moving angle (in the initial position) is equal to the angle between the interval and the directrix (both on the same side of the interval); this curve turns out to be a parabola. A natural question is, What happens if these angles are different? In chapter I Jan de Witt indeed raises this question. He says, however, that it would be possible to investigate this possibility, but that he prefers another method, namely the construction of a curve, using a rotating straight line and a sliding angle, instead of a rotating angle and a moving straight line. Chapter II is devoted to this construction. In this chapter, IV, he returns to this question and he shows that the method of chapter I, applied to the case of different angles, leads to the construction of a hyperbola.

As an "extra" he discusses three special cases that can occur in constructing a hyperbola:

i. The tangent at the vertex is perpendicular to an asymptote.
ii. The moving angles are right angles.
iii. The generator in the initial position is perpendicular to the directrix.

C. In this section he gives two other methods to construct an ellipse and makes a final remark.

i. Two mutually perpendicular straight lines (EMF and CMD) pass through the point M (see Figure 2.25). The line segment AMB (AM not necessarily equal to MB) rotates about the point M. Through A and B the straight lines l and m are drawn parallel to EMF and CMD, respectively. Now we are interested in the curve that is described by the point of intersection (S) of the moving lines l and m when AMB rotates about M. This curve turns out to be an ellipse with center M whose axes lie along CMD and EMF. Jan de Witt remarks that this curve is a circle if $AM = MB$.

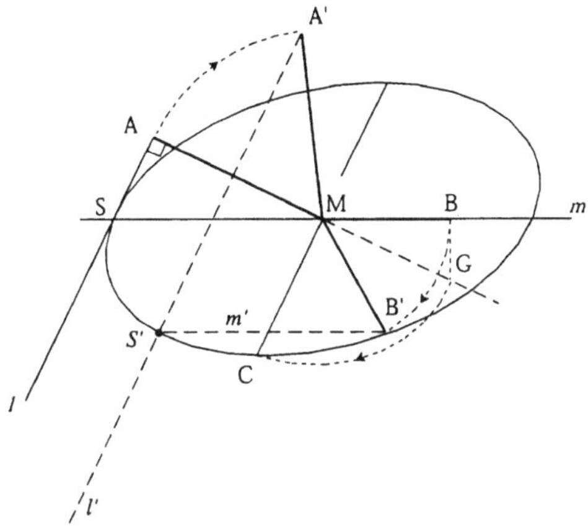

FIGURE 2.26

ii. Next the case is considered in which the angle AMB is not a straight angle (see Figure 2.26). In this case the curve is generated as follows: a straight line l is drawn through A perpendicular to MA intersecting the line m through B and M at S. Then the angle AMB rotates about M and in the new position $A'MB'$ of the angle AMB, the lines l' and m' are drawn through A and B parallel to l and m, respectively. The point of intersection of l' and m' is called S' and we are interested in the curve described by S', if AMB continuously rotates about the point M and if in every new position of the angle AMB lines are drawn parallel to l and m. This curve turns out to be an ellipse with M as its center and MS as a semidiameter. The corresponding second diameter is parallel to AS; its length is twice that of the segment MG, whereby G has been selected on the produced part of AM, in such a way that its projection on m coincides with the point B. Jan de Witt remarks that the curve described by S consists of two straight lines parallel to m if the angle AMB is a right angle.

iii. Finally de Witt remarks that a method analogous to that explained in the first part of chapter III to describe an ellipse might also be used to generate a hyperbola or a parabola. In his opinion, however, he has said enough about the generation of parabolas, hyperbolas, and ellipses, and he switches to the next topic:

Locorum Planorum, Solidorumque inventiones ac determinationes,

i.e., the finding and determining of plane and solid loci, which is the subject of *Elementa Curvarum Linearum, Liber Secundus.* The central theme of this book is the treatment of the straight line, the parabola, the hyperbola, and the ellipse by means of their algebraic equations.

3

Latin text and translation

Jan de Witt

Foundations of Curved Lines

published in collaboration with

Frans van Schooten
Professor of Mathematics
at Leiden University

Amsterdam
Printed by Blaeu, MDCLXXXIII
financed by the Company

JOHANNIS DE WITT

ELEMENTA

CVRVARVM LINEARVM.

Edita

Operâ FRANCISCI à SCHOOTEN,
in Academia Lugduno-Batava Matheſeos
Profeſſoris.

AMSTELODAMI,
Ex Typographia BLAVIANA, MDCLXXXIII.
Sumptibus Societatis.

The most Illustrious and most Learned Sir FRANCISCUS van SCHOOTEN is respectfully greeted by JOHANNES de WITT

The Ancients have, at least in my opinion, passed down in a sufficiently clear way the true and essential nature and the principal properties of the straight lines, the angles enclosed by them, the rectilineal figures that arise from them, and the circles.

Those who followed showed more extensively and more clearly how from these traditions, indeed from a few and fundamental principles thereof, arbitrary plane* problems can be solved and disentangled. In general, they showed that all problems that one could possibly want to pose, when contemplating and studying straight lines, angles, rectilineal figures, and circles can be solved.

They did this by forming and solving equations using a certain analytical method. Indeed, if only one circle has been given, be it very small or enormous, then anyone, even if only moderately conversant with the said precepts of the ancient and modern geometers, can very easily solve arbitrary plane problems by means of straight lines alone.**

Clariſſimo, Doctiſſimoque Viro,

D°. FRANCISCO à SCHOOTEN,

IOHANNES DE WITT

S. P. D.

LInearum rectarum, angulorumque, quos comprehendunt, ut & figurarum rectilinearum, quæ inde naſcuntur, nec non Circulorum naturam veram atque intrinſecam, proprietateſque præcipuas, meo quidem judicio, ſatis perſpicuè tradiderunt Antiqui, ac quo pacto ex iiſdem traditis, imò ex paucis & principalioribus eorundem principiis, quælibet Problemata Plana, ac generaliter quæcunque in linearum rectarum, angulorum, figurarumque rectilinearum, nec non Circulorum contemplatione & cognitione deſiderari queunt, reſolvantur atque eruantur, univerſali quâdam viâ & Methodo Analyticâ, per Æquationum inventionem, harumque reſolutionem, pleniùs planiùſque à Recentioribus oſtenſum eſt; Adeò ut vel unico Circulo dato, utut exiguo aut ingenti, quæcunque Problemata Plana per ſolas lineas rectas unuſquiſque, in dictis Antiquorum Recentiorumque Geometrarum præceptis mediocriter verſatus, facillimè reſolvat; ac pro-

I therefore always thought it superfluous and senseless to desire more or different propositions and proofs in this field.

However, when I more carefully studied the textbooks on the other curved lines–as far as they have been transmitted by the Ancients and explained by the later mathematicians–I thought it absolutely against the natural order, which in mathematics one should respect as much as possible, to seek the origin of these curves in a solid and then to transfer them to the plane.

For the same reason, and because of the various complicated reasonings upon which they often rest, I found the proofs explained in these textbooks rather difficult to understand in many places. The long chain of propositions makes the proofs tedious for the readers and a burden for the memory. Stimulated, long ago, by this consideration, I devoted my spare time to the study of the Humaniora and the other Liberal Arts. I observed that curves of any genus [1.20], not only those curves that are commonly called conic sections, surely arise from various intersections of solids that are differently composed or shaped, but can be generated separately, in an infinite number of ways, in the plane as well. Moreover, their nature and their properties can be derived much more easily from this way of producing them than from the intersection of solids.

inde de iiſdem vel plura vel alio modo propoſita ac demonſtrata quædam deſiderare, & ſupervacuum & ineptum ſemper exiſtimavi. At verò cum cæterarum linearum curvarum Elementa, prout à Veteribus tradita atque à Recentioribus explicata ſunt, diligentiùs conſideraſſem, originem earum è ſolido peti atque inde ipſas in planum transferri naturali ordini, qui in Mathematicis quàm maximè obſervandus eſt, omnino contrarium duxi; quemadmodum & demonſtrationes in iiſdem Elementis propoſitas, multis in locis eadem de cauſa & propter varias rationum compoſitiones, quibus ſæpe innituntur, ſubobſcuras, ac longa Propoſitionum ſerie Lectoribus tædio memoriæque oneri eſſe judicavi. Atque eâ quidem contemplatione excitatus jampridem, dum ſtudiis humanioribus Liberaliumque Artium doctrinæ incumbere mihi otium erat, animadverti, non eas ſolùm, quas vulgò Coni ſectiones *appellarunt, ſed & omnes omnino curvas lineas, cujuſcunque ſint generis, multipliciter quidem ex varia corporum diverſimodè compoſitorum aut figuratorum ſectione gigni, at verò earundem ſingulas infinitis quoque modis in plano generari, ipſarum autem naturam & proprietates ex ea generatione multò faciliùs quàm ex corporum ſectione deduci, ac firmiter mihi perſuaſum habeo, nullam aliam eſſe cauſam, quod*

linea-

Nobody before this time had explained and proved the origins, the nature, the properties, and the essence, nor given an accurate enumeration of the species, of the curves of the second genus and higher degrees [1.20].

This true, I am convinced, for no other reason than that, both when treating the origins and the production of the curves belonging to the first genus and when proving their essence and their properties, reasoning had been deflected from the most natural and simplest way.

Indeed, if one considers them just as they are generated in the plane, in the simplest and yet different ways, then understanding and the imagination are guided towards the generation of the curves of the second genus, as it were, spontaneously.

As you, my dear van Schooten, now desire to have access to the writings related to these topics, which I studied in the past when I had the spare time to make this possible, well, I will comply with your wish, so far as I am able. I will now send you my earlier writings on this subject, which I found in fairly good order and which I put wholly at your disposal.

However, I will gather the rest, which is scattered around, at least if more important business will allow this, and put it together in an orderly manner. I will also send this to you in due course when it is gathered and well ordered.

Farewell.

The Hague, October 8, 1658.

linearum curvarum secundi generis ulteriorumque graduum ortus, natura, proprietas, atque essentia, cum exacta specierum enumeratione, à nemine antehac explicata ac demonstrata sint, quam quòd tam in tractatione ortus & generationis, quàm in demonstratione essentiæ ac proprietatum linearum curvarum primi generis à naturali & simplicissima via deflexum sit, utpote cum earundem contemplatio, prout in plano simplicissimè & quidem diversimodè generantur, intellectum & imaginationem ad genesin linearum curvarum secundi generis quasi sponte ducat. Cumque eorum, quæ antehac, dum per otium licuit, eò spectantia meditatus sum, tu nunc amicissime Schooteni, *copiam tibi fieri desideres, en, quantum in me est, desiderio tuo satisfacio, quæque de eodem argumento a me quondam conscripta ac pene in ordinem redacta inveni, jam tibi mitto, tuique omnino juris facio, cætera autem, quæ sparsim tantùm annotata sunt, si modò graviora id ferent negotia, recolligam, debitoque ordine conjungam; recollecta, atque ordinata suo quoque tempore tibi missurus, Vale. Hagæ Com.* VIII *Octobr. Anni* M. DC. LVIII.

Notes

* Plane problems (*problemata plana*) are problems that can be solved by using a ruler and a compass alone. In addition to these, there are the so-called solid problems (*problemata solida*), the solution of which is possible only by means of conic sections. Finally, there are linear problems (*problemata linearia*) which can only be solved by means of "higher" curves, such as spirals, quadratrices, conchoids, etc.

** The background of this remark is probably the following: the ancient geometers wondered what geometrical constructions could be performed using a ruler and a compass of fixed opening. The reason for the presentation of this problem might be the idea that these constructions should be more accurate. The first mathematician who solved a number of problems in this way, was Abu"l-Wafa" (940–997/998). Battista Benedetti (1530–1590), in his *Resolutio* (1552), gave the general solution of all the construction problems in Euclid"s *Elements*, using only a ruler and a compass of fixed opening. Tartaglia (1499–1557) and Ferrari (1522–1565) occupied themselves with these problems as well. Georg Mohr (1640–1697) showed in his *Euclides Danicus* (1672) that every construction that can be performed with ruler and compass can be performed with only a compass as well. Perhaps he was inspired by Benedetti"s work. Of course, it is to be understood that this concerns the construction of points. Abu"l-Wafa" and Benedetti could construct only one circle and Mohr could not draw a straight line!

Another generalisation was given by Poncelet (1788–1867) in his *Traité des Propriétés Projectives des Figures* (1822). In this book he showed that all constructions that can be performed with a ruler and a compass can be performed by using a ruler alone, if in the plane one fixed circle is given, together with its center (of course, with the same restrictions we mentioned before). Jakob Steiner (1769–1863) published a more elegant proof of this theorem in his book *Die Geometrischen Konstruktionen ausgeführt mittelst der geraden Linie und eines festen Kreises, als Lehrgegenstand auf höheren Unterrichtsanstalten und zur praktischen Benützung.*

Probably de Witt's remark refers to what was proved by Benedetti; in my opinion the expression *unico circulo dato* refers to a compass of fixed opening, which determines a unique circle. Or should de Witt have anticipated the results of Poncelet and Steiner?

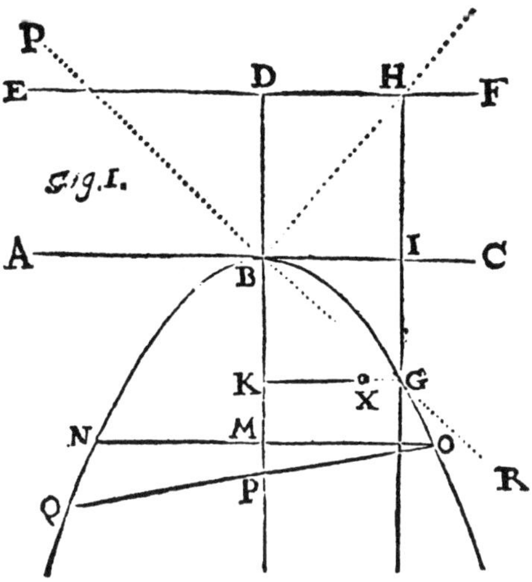
Fig. I.
P
E
D
H
F
A
B
I
C
K
X
G
N
M
O
R
Q
P

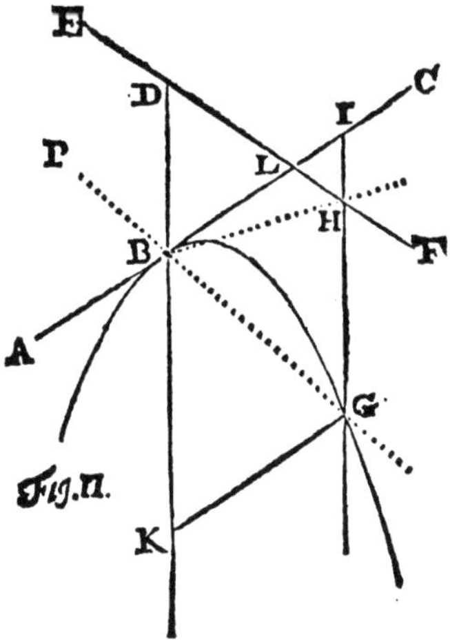
Fig. II.
E
D
I
C
P
L
H
F
B
A
G
K

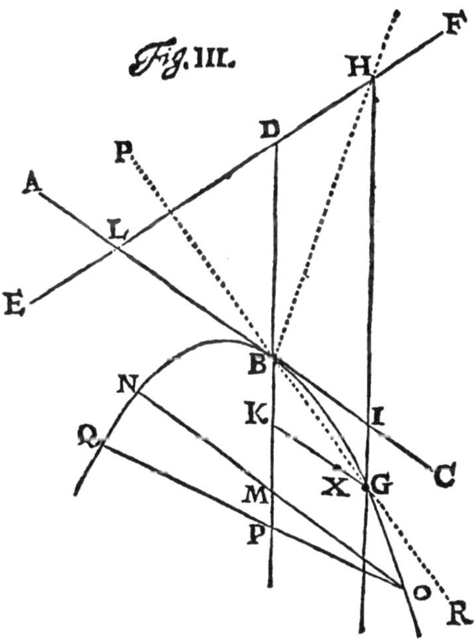
Fig. III.
F
H
D
P
A
L
E
B
N
K
I
Q
X
G
C
M
P
O
R

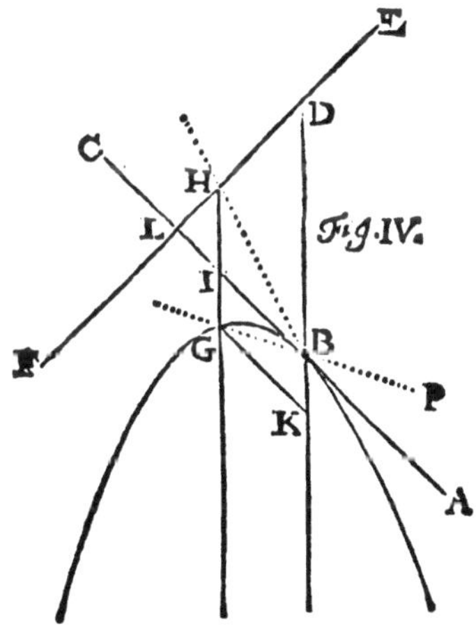
E
D
C
H
L
Fig. IV.
I
G
B
F
P
K
A

[159]

Jan de Witt

Foundations of Curved Lines

First Book

Chapter I

First Definitions

I

When along a fixed straight line another straight line moves by means of a fixed point on it, such that this other straight line is always parallel to itself, or coincides with itself [1.1], and when during this same motion an angle [1.2] rotates about a fixed point (coinciding with its vertex) such that one arm of this angle always passes through the aforementioned moving point and at the same time the point of intersection of the other arm with the aforementioned moving straight line describes a curve, then the straight line, that–as is said–always moves parallel to itself or coincides with itself, will be called the generator [1.3].

II

The other straight line, however, that remains fixed, will be called the directrix.

III

The aforementioned angle and its adjacent angle will appear under the name moving angles.

JOHANNIS DE WITT ELEMENTA CVRVARVM LINEARVM.

LIBER PRIMVS.

CAPUT I.

DEFINITIONES PRIMÆ.

I.

SI per rectam lineam immotam altera recta certo ſui puncto ſibi ſemper parallela moveatur aut incedat, eodemque illo motu anguli cujuſdam rectilinei, circa punctum fixum (quod idem ſit cum ejus vertice) circulariter mobilis, crus unum ſemper per prædictum mobile punctum tranſiens ſecum ducat, atque ita ſimul cruris alterius, & dictæ lineæ incedentis interſectione curva deſcribatur linea; recta, quæ, uti prædictum eſt, ſibi ſemper parallela movetur aut incedit, *Deſcribens* dicetur.

II.

Altera verò recta, immota manens, *Directrix* vocabitur.

III.

Prædictus autem angulus rectilineus, atque is qui ei eſt deinceps, *Angulorum mobilium* nomine venient.

IV.

[160]

IV

The angles between the generator and the directrix will be called the angles with the directrix.

V

The fixed point about which the moving angle rotates will be called the pole.

VI

The part of the generator cut off between the pole and the directrix [1.4] will be called the interval.

VII

The arm of the moving angle that is carried along by the generator will be called the passive arm.

VIII

The other arm, which is intersected by the generator, will be called the active arm and, after extension beyond the vertex of the angle, will be called the active line.

IX

When the generator passes through the pole and by consequence coincides with the passive arm, then we say that both the generator and the passive arm, as well as the active line and the whole moving angle, are in the initial position; when they are referred to in this way, we assume that they are in this position [1.5].

X

We shall say that any curve generated in the plane by the intersection, as mentioned before, has been generated by the active line and the interval, considered as they appear and mutually linked in the initial position. In this position, the interval coincides with the generator as well as with the passive arm, and the active line encloses, with the interval, the moving angle on both sides.

IV.

At quos *describens* ad *directricem* efficit, *Anguli ad Directricem* dicentur.

V.

Punctum fixum, circa quod *angulus mobilis* circulariter movetur, *Polus* nuncupabitur.

VI.

Ea autem *describentis* pars, quæ inter *Polum* & *directricem* intercipitur, *Intervallum* nominabitur.

VII.

Crus *anguli mobilis*, quod *describens* secum ducit, *Crus Patiens*.

VIII.

Alterum verò crus, quod à *describente* secatur, *Crus Efficiens*, & per anguli verticem productum, *Linea Efficiens* appellabitur.

IX.

Cum *describens* per *Polum* transit, ac proinde & cum *crure patiente* coïncidit, esse tam *describentem* quàm *crus patiens*, ut & *lineam efficientem* totumque *angulum mobilem* in *statione prima* constitutum dicemus; ac quoties de iis simpliciter sermo erit in tali ipsas positione considerabimus,

X.

Quamlibet curvam, intersectione, uti prædictum est, in plano genitam, descriptam dicemus, *efficiente* atque *intervallo* considerati, ut exhibentur ac sibi invicem junguntur *in statione prima*; adeò ut *efficiens* cum *intervallo*, quod tam cum ipsa *describente* quàm cum *crure patiente* in eadem statione coïncidit, *angulum mobilem* utrinque constituat.

Ut

[161] We refer to the added figures. Let us imagine that the straight line HG always moves parallel to itself by means of one of its fixed points, e.g., H, along a fixed straight line EF and that during this same motion it carries along the arm BH of the angle HBG,

Vt in appositis figuris, si recta HG sibi semper parallela certo sui puncto, puta H, moveri concipiatur per immotam EF, eodemque illo motu secum ducere crus BH anguli HBG, circu-

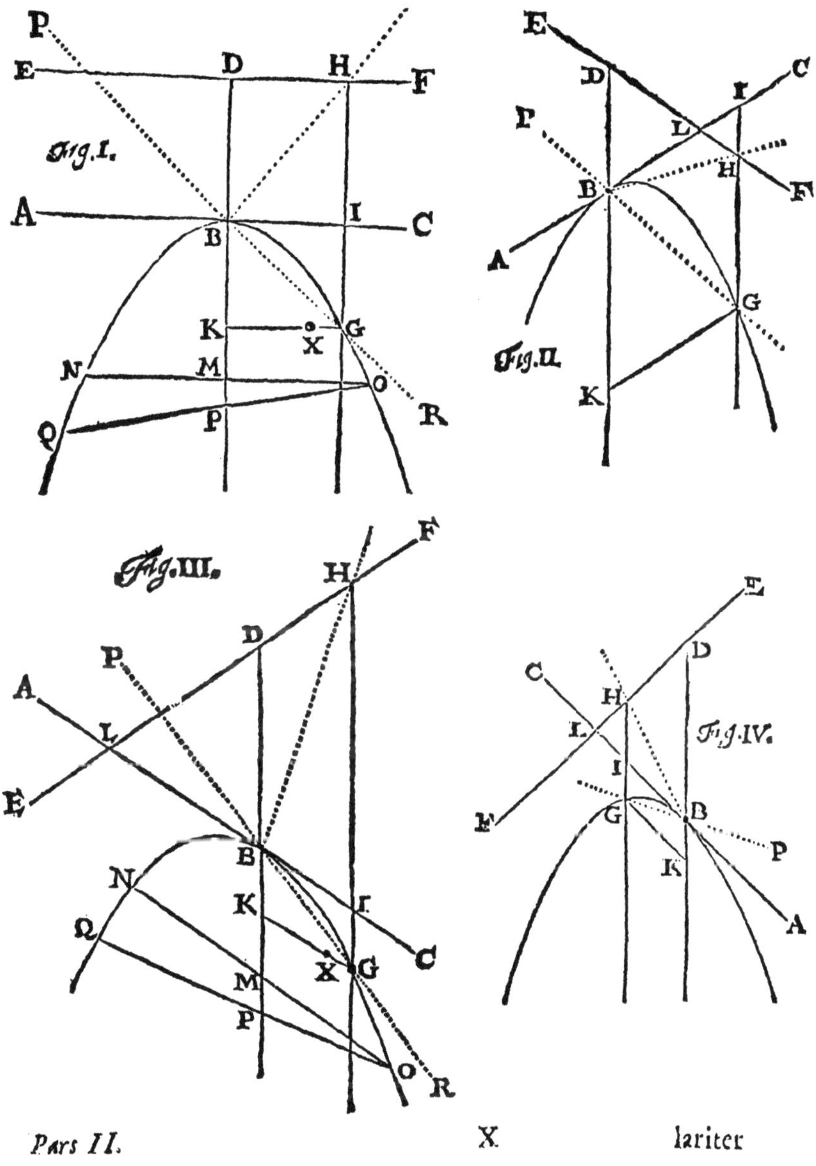

[162] which rotates about the point B such that the arm BH always passes through the aforementioned point H on HG, and so that at the same time the point of intersection G of the other arm BG with the aforementioned straight line HG describes the curve BG.

Then the following will be true:

HG is the generator.

EF is the directrix.

HBG and HBP are the moving angles.

FHG and EHG are the angles with the directrix.

B is the pole.

BD is the interval.

BH is the passive arm.

BG is the active arm.

PG is the active line.

DK is the generator in the initial position, or simply the generator.

DBC and DBA are the moving angles in the initial position.

AC is the active line in the initial position, or simply the active line.

We shall say that the curve BG has been generated by the active line AC but with the interval BD. It is also clear that the passive arm BH coincides with the interval BD when the active line PG has the same position as AC, that in this case the generator HG is in the same position as DK, and that the active line and the interval on both sides include the moving angles DBC and DBA.

Theorem I

Proposition 1

If a curve has been generated with any active line and any interval so that the moving angles are equal to the angles with the directrix on the same side of the interval, then this curve will have the following property: The square on the line segment parallel to the active line (in the initial position), drawn from an arbitrary point on the curve to the generator (in the initial position), has the same area as the rectangle included by the interval and that part of the generator that is cut off between the pole and the applied line segment [1.6].

Let the curve BG be generated by the active line ABC, the interval BD, and the directrix EF, so that the moving angle DBA equals the angle EDB with the directrix, and let the line segment GK be drawn from a point G (chosen at random on the curve) to the generator DBK, parallel to the active line AC (in the initial position, transl.); then I say that the square on the applied line segment equals the rectangle DBK [1.7].

lariter mobilis circa punctum B; ita ut idem crus BH ſemper tranſeat per prædictum ipſius HG punctum H, ſimulque alterius cruris BG ac dictæ lineæ HG interſectione G deſcribatur curva linea BG: erunt

HG *deſcribens.*
EF *directrix.*
HBG, HBP *anguli mobiles.*
FHG, EHG *anguli ad directricem.*
B *Polus.*
BD *intervallum.*
BH *crus patiens.*
BG *crus efficiens.*
PG *linea efficiens.*
DK *deſcribens in ſtatione prima*, ſive *deſcribens* ſimpliciter.
DBC, DBA *anguli mobiles in ſtatione prima.*
AC *efficiens in ſtatione prima*, ſive *efficiens* ſimpliciter.

Curvam BG, *efficiente* AC, *intervallo* verò BD deſcriptam dicemus; Et apparet, cùm *efficiens* PG eſt in ſtatione AC, *crus patiens* BH coïncidere cum *intervallo* BD; ac *deſcribentem* HG tunc eſſe in ſtatione DK, atque per *efficientem* & *intervallum* conſtitui utrinque *angulos mobiles* DBC, DBA.

THEOREMA I.

Propoſitio 1.

Quâlibet *efficiente*, & quocunque *intervallo*, ſi *anguli mobiles* æquales ſint iis, qui *ad directricem* ſunt ab eadem parte, curvâ deſcriptâ, hoc ipſi proprium erit, ut quævis recta à quolibet curvæ puncto ad *deſcribentem efficienti* æquidiſtans applicata poſſit rectangulum, ſub *intervallo* atque ea *deſcribentis* parte, quæ inter *Polum* & applicatam intercipitur, contentum.

Sit *efficiente* ABC, *intervallo* BD, & *directrice* EF deſcripta curva BG; ita ut *angulus mobilis* DBA ſit æqualis angulo EDB *ad directricem*, ſitque à puncto G in curva utcunque aſſumpto ad *deſcribentem* DBK applicata recta GK *efficienti* AC parallela: dico quadratum applicatæ GK rectangulo DBK æquale eſſe.

Con-

[163] Indeed, let the moving angle as well as the generator be brought into the position in which the point G was generated by their intersection, e.g., HBG and HIG; if the moving angle as well as the angle

Constitutis enim tam *angulo mobili* quàm *describente* in statione uti fuêre, cùm per ipsarum intersectionem descriptum est punctum G, veluti in HBG & HIG: si tam *angulus mobilis* quàm is

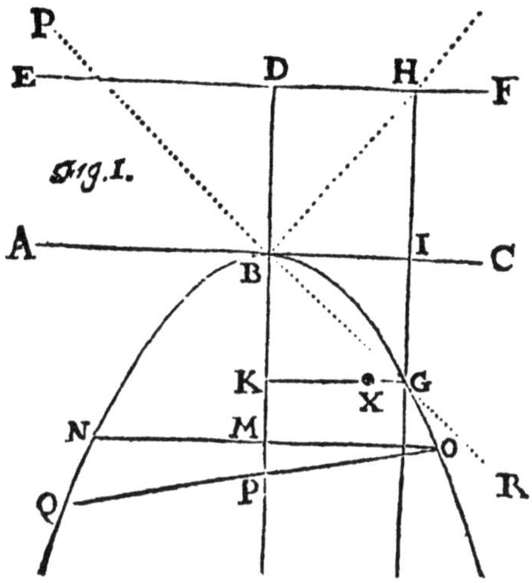

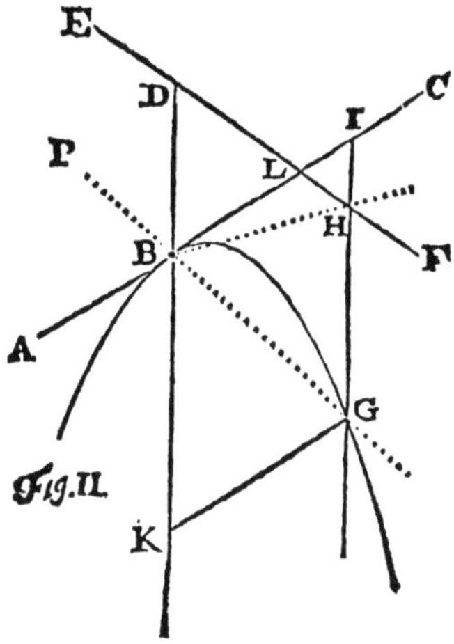

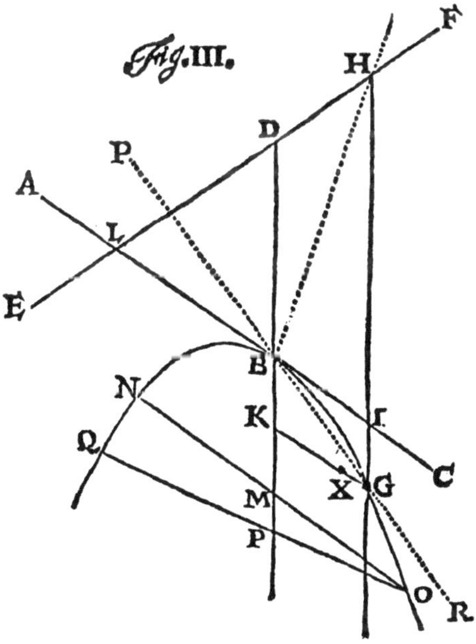

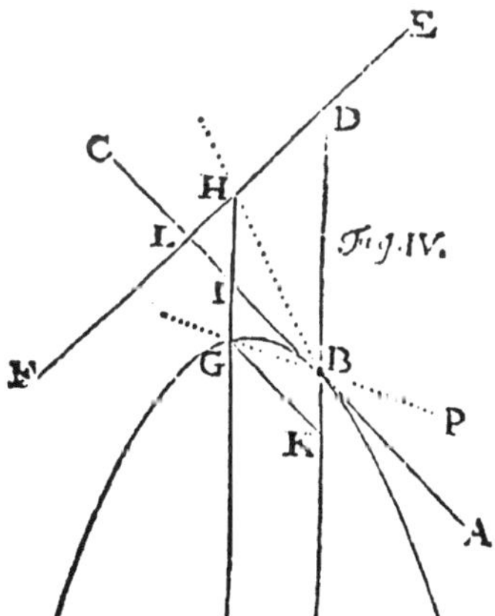

qui-

[164] with the directrix is a right angle, as in the first figure (page [163], transl.), then[1] IB will be to IG as HI is to IB, i.e.,[2] GK is to BK is as DB to GK and therefore[3] the square on the line segment GK equals the rectangle DBK.

If, however, each of the angles ABD and EDB is oblique, as in the remaining figures, the active line and the directrix will, being produced, intersect at the side (of the interval, transl.) where the moving angle and the angle with the directrix are acute. Suppose they meet each other at the point L. Now the angles LBD and LDB are equal (by construction), as are LIH and LHI (because DB and HI are parallel);[4] therefore[5] the line segments LD and LB as well as LH and LI are also equal, so their sums[a] as well as their differences[b] DH and BI will be equal. The angle DBH, however, equals the angle IBG, i.e.,[6] BGK; this is obvious when we add[c] to or subtract[d] from, the identical or equal angles DBI and HBG the common angle HBI. The angle BDH equals the angle DBI, i.e.,[7] BKG by construction. Therefore[8] the triangles BDH and GKB have the same angles, and so[9] GK is to KB as BD is to DH or BI, i.e.[10], to GK. Therefore, as above[11], the square on the applied line segment GK equals the rectangle DBK. This is the statement. So it is clear that the curve, generated by the aforementioned intersection, is the same as the one the classical mathematicians called a parabola; the pole is the point they called a vertex; the generator in the initial position is the same as the diameter or–if the moving angles are right angles–the axis; the interval is what they called the latus rectum–called the parameter by modern writers– belonging to this diameter or axis, and the lines parallel to the active line (in the initial position, transl.) are those which they called ordinate-wise applied to the diameter or axis. Therefore we have to adhere to the same names [1.9].

Corollary 1

As the generator and the active line in arbitrary position intersect at just one point, it is clear that the generator in an arbitrary position

---

[1]VI, 8, Cor.; [2]I, 34; [3]VI, 17; [4]I, 29; [5]I, 6; [a]in the case of Fig. II and the like; [b]in the case of Figs. III and IV and the like; [c]in the case of Fig. II and the like; [d]in the case of Figs. III and IV and the like; [6]I, 29; [7]I, 29; [8]I, 32; [9] VI, 4; [10]I, 34; [11]VI, 17.

qui ad *directricem* est rectus sit, uti in prima figura, erit [1] ut HI ad IB, ita IB ad IG, id est [2], ut DB ad GK, ita GK ad BK. ac proinde [3] quadratum rectæ GK rectangulo DBK æquale erit.

At verò si obliquus fuerit uterque angulorum ABD, EDB, uti in cæteris figuris, secabunt sese *efficiens* & *directrix* productæ ad eas partes, ubi *angulus mobilis*, isque qui ad *directricem* est, acuti erunt. sit itaque ipsarum intersectio in L puncto. Quoniam igitur tam anguli LBD, LDB, ex constructione, quàm LIH, LHI, propter parallelas DB, HI, æquales sunt [4]; erunt quoque [5] tam lineæ LD, LB, quàm LH, LI; ac proinde & compositæ [a] vel residuæ [b] DH, BI æquales. Cum autem, angulis DBI, HBG iisdem, sive æqualibus existentibus, addito [c], vel ablato [d] communi angulo HBI, angulus DBH angulo IBG, id est [6], BGK, fiat æqualis, atque angulus BDH ex constructione angulo DBI, id est [7], BKG, sit æqualis: erunt [8] triangula BDH, GKB æquiangula, eritque proinde [9], ut BD ad DH sive BI, hoc est [10], ad GK, ita eadem GK ad KB. quare, ut supra [11], quadratum applicatæ GK rectangulo DBK æquale erit. Quod est propositum.

1 per Cor. 8 *sexti Eucl.* 2 *per* 34 *primi.* 3 *per* 17 *sexti.* 4 *per* 29 *primi.* 5 *per* 6 *primi.* a in casu fig. II & similibus. b in casib. fig. III & IV, aut similibus. c in casu fig. II. & sim. d in casib. fig. III & IV, aut similibus. 6 *per* 29 *primi.* 7 *per* 29 *primi.* 8 *per* 32 *primi.* 9 *per* 4 *sexti.* 10 *per* 34 *primi.* 11 *per* 17 *sexti.*

Constat itaque, curvam intersectione, uti prædictum est, descriptam, eam ipsam esse, quæ Veteribus Parabola; *Polumque* idem punctum quod vertex; *lineam* autem *describentem in statione prima* eandem quæ diameter, aut si *anguli mobiles* recti fuerint, quæ axis; *intervallum* verò idem quod latus rectum sive recentioribus Parameter ad eandem diametrum eundemvè axem pertinens; atque *efficienti* parallelás, eas, quæ ordinatim ad diametrum vel axem applicatæ dicebantur; quare & eadem nomina retinento.

Corollarium I.

Cum *describentis efficientisque* intersectio quibuscunque stationibus in uno tantùm puncto fiat, manifestum est, *describentem* in quacunque

[165] intersects the parabola at just one point, which means that all straight lines parallel to a diameter meet the parabola at just one point.

Corollary 2

When the generator recedes continuously from the pole, the angle between the active arm and the active line in the initial position, e.g., GBI, grows steadily. Therefore it is clear that an arbitrary line segment, drawn from the pole to any point on the curve, e.g., BG, lies totally inside the parabola; however, its extension, e.g., to R, lies outside the parabola.

Corollary 3

Moreover, it is clear that the angle GBK may be reduced boundlessly and can be made smaller than any given angle, but the active arm BG can never coincide with the generator BK, let alone intersect it: therefore it would be

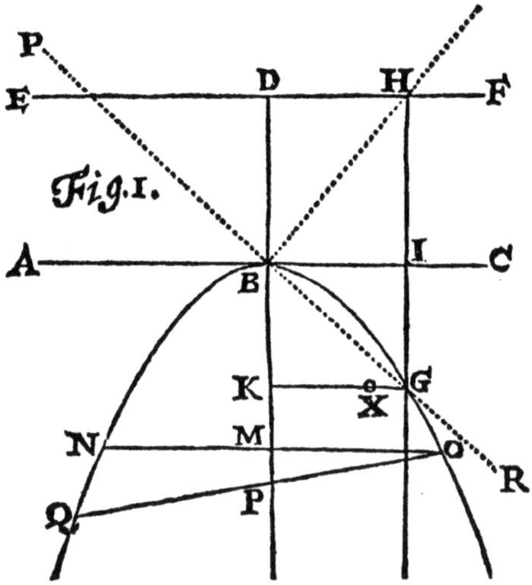

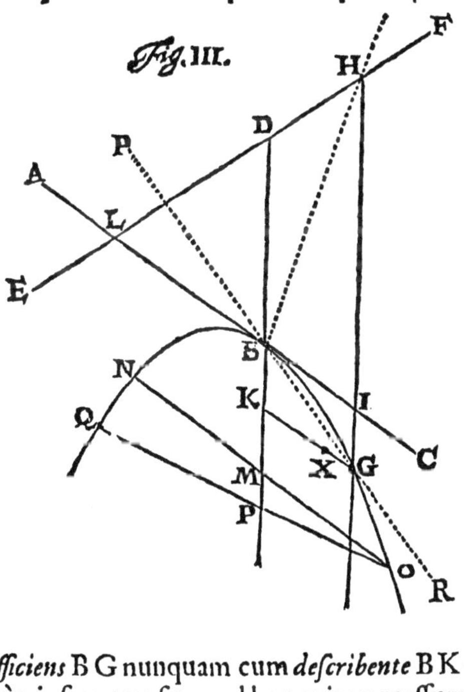

cunque ſtatione, id eſt, rectas omnes diametro æquidiſtantes, in uno tantùm puncto Parabolæ occurrere.

Corollarium 2.

Cumque continuo *deſcribentis* à *Polo* receſſu major majorque ſemper fiat angulus, quem *crus efficiens* conſtituit ad *lineam efficientem in ſtatione prima*, veluti G B I, manifeſtum eſt, quamlibet rectam à *Polo* ad quodlibet curvæ punctum ductam, ut, ex. gr., B G, totam intra Parabolam, productam autem, uti ad R, extra Parabolam cadere.

Corollarium 3.

Conſtat præterea angulum G B K indefinitè quidem diminui, omnique propoſito angulo rectilineo minorem reddi poſſe; ſed *crus* tamen *efficiens* B G nunquam cum *deſcribente* B K coincidere, multò minùs ipſam tranſire: ad hoc enim neceſſum foret, ut *crus patiens*

[166] necessary for the passive arm BH to be parallel to the generator EF[1] or at least to lie below the straight line drawn from the pole parallel to the directrix, which is clearly impossible as it (i.e., the passive arm, transl.) always intersects the directrix.

Corollary 4

Therefore it is also clear that all straight lines that intersect the axis or a diameter of the parabola, if produced, finally meet the parabola. Indeed, let the straight line KX intersect the diameter BKM and let the active arm BG–being in such a position that the angle KBG is smaller than a given angle MKX[2]–intersect the parabola at the point G as well.[3] Since the straight line KX intersects the arm BG, the point of intersection will lie between B and G, in which case the straight line being produced will intersect the curve BG, or it will intersect this arm at the point G itself, in which case this straight line intersects the curve at this point; or, finally, it will intersect BG beyond G, in which case it intersects the parabola anyway.

Corollary 5

Moreover, it is clear that all line segments that have been applied ordinate-wise [1.10] and that are bounded by the parabola on both sides, are bisected by the axis or by the diameter. Indeed, if NMO has been applied ordinate-wise, then[5] the square on NM as well as the square on MO equals the rectangle DBM; therefore these squares are mutually equal and so the line segments NM and MO are also equal.

Corollary 6

The converse of the preceding is also clear: apart from the line segments that are parallel to the active line (in the initial position, transl.), no other line segments can be bisected in a parabola by the axis or by the diameter. Indeed let OQ–not parallel to AC–be bisected by the axis or by the diameter BP in P. If one should draw ON parallel to the active line AC (ON therefore would be bisected by the same axis or diameter at M[6]), then OP would be to PQ as OM is to MN and therefore[7] the line through N and Q would be parallel to the diameter and would meet the parabola at the two points N and Q, which is impossible.[8] Therefore not only are all line segments that have been ordinate-wise applied bisected by the diameter, but also all line segments that are bisected by a diameter have been applied ordinate-wise

---

[1]I, 29; [2]because of the preceding Cor.; [3] Cor. 2, p. [165]; [4]Cor. 2, p. [165]; [5]Prop. 1, p. [162]; [6]according to the preceding Cor.; [7]VI, 2; [8] Cor. 1 p. [164].

B H *directrici* E F foret parallelum [1], aut certè ut caderet infra eam, quæ à *Polo directrici* æquidiſtans ducta eſſet, quod planè impoſſibile eſt, cum *directricem* ſemper ſecet.

[1] per 29 primi.

Corollarium 4.

Ideoque apparet, rectas omnes, quæ Parabolæ axem vel diametrum ſecant, productas tandem Parabolæ occurrere. Secet enim recta K X diametrum B K M, ac *crus efficiens* B G, in ea ſtatione conſtitutum, ut K B G angulus minor ſit dato angulo M K X [2], per Parabolam tranſeat in puncto G. Quoniam igitur recta K X cruri B G occurrit, aut eidem occurret inter B & G, quo caſu ipſa producta etiam curvæ B G occurſura eſt [3], aut eidem in ipſo G puncto occurret, quo caſu & ſimul Parabolæ ibidem occurret, aut denique ipſi B G occurret ad partes G productæ, quo utique caſu priùs Parabolæ occurret [4].

[2] juxta Cor. præcedens.
[3] per Corol. 2 hujus.
[4] per Cor. 2 hujus.

Corollarium 5.

Manifeſtum quoque eſt, applicatas omnes, utrinque Parabolâ terminatas, ab axe aut diametro bifariam dividi. Vt, ſi ducta ſit applicata N M O, quoniam [5] tam quadratum N M quàm quadratum M O æquale eſt rectangulo D B M: erunt quoque eadem quadrata inter ſe æqualia, ac proinde & rectæ N M, M O æquales.

[5] per 1 hujus.

Corollarium 6.

Patet quoque præcedentis converſum, nempe non poſſe alias rectas præter eas, quæ *efficienti* æquidiſtant, in Parabola ab axe ſive diametro bifariam ſecari. Si enim O Q, quæ non ſit æquidiſtans ipſi A C, ab axe ſive diametro B P bifariam divideretur in P, ductâ O N *efficienti* A C parallelâ, quæque proinde ab eodem axe ſive diametro bifariam quoque ſecabitur in M [6], foret O P ad P Q, ut O M ad M N: ideoque [7] ducta recta per N & Q eſſet diametro parallela, ac Parabolæ occurreret in duobus punctis N & Q. quod fieri non poteſt [8].

Itaque non ſolùm applicatæ omnes à diametro bifariam dividuntur, ſed & quæ à diametro biſecantur ad ean-

[6] per Corol. præcedens.
[7] per 2 ſexti.
[8] juxta Corol. 1 hujus.

[167] to this diameter, and if a diameter bisects an arbitrary chord in a parabola, it will bisect all line segments parallel to it as well.

Corollary 7

From what has been proved it follows easily that the squares on the ordinate-wise applied line segments are in the same ratio as the parts of the diameter cut off between the vertex and the applied line segments.

Let GK and NM be applied ordinate-wise; then[1] the square on the line segment GK will be to the square on NM as the rectangle DBK is to the rectangle DBM, that is[2] as BK is to BM.

Corollary 8

From the way in which the curve is generated itself, it is furthermore obvious that the active line in the initial position, that is, the straight line drawn through the pole or vertex parallel to the ordinate-wise applied chords, touches the parabola there and at no other point. It would certainly not intersect the parabola.

[1]Prop. 1, p. [162]; [2]VI, 1.

eandem ordinatim applicatæ ſunt: & ſi diameter rectam quamlibet in Parabola ductam biſariam dividat omnes quoque ipſi æquidiſtantes biſariam ſecabit.

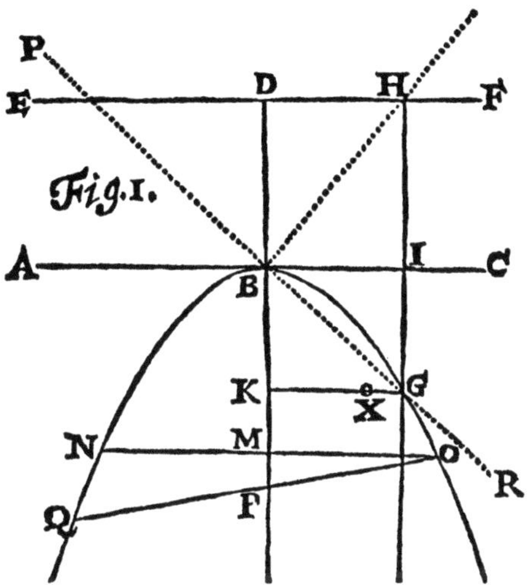

Fig. I.

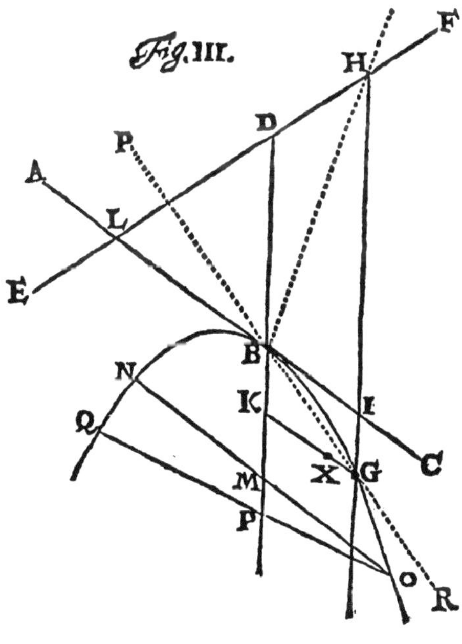

Fig. III.

Corollarium 7.

Ex demonſtratis quoque facilè colligitur, applicatarum quadrata ad ſe invicem eſſe, ſicut ad ſe invicem ſunt diametri portiones inter verticem & applicatas interceptæ. Vt, ſi applicatæ ſint G K, N M, erit [1] quadratum rectæ G K ad quadratum ipſius N M, ut rectangulum D B K ad rectangulum D B M, id eſt [2], ut B K ad B M.

[1] *per 1 hujus.*

[2] *per 1 ſexti.*

Corollarium 8.

Ex ipſa porrò deſcriptione manifeſtum eſt, *efficientem in ſtatione prima*, id eſt, rectam, quæ per *Polum* ſive verticem applicatis æquidiſtans ducitur, ibidem Parabolam nec in alio præterea puncto contingere, multò minùs eandem ſecare.

[168] Indeed, if on the curve an arbitrary point is chosen, distinct from the pole B, e.g., G, and if the active arm is applied through this point, e.g., in the position BG, then there is an angle formed by this arm and the active line (in the initial position, transl.), let us say GBC, and therefore the arbitrarily chosen point G, i.e., the whole parabola with the exception of the pole B, will lie below the active line ABC.

Corollary 9

From the preceding it is also clear that apart from the active line no other straight line can touch the parabola at the pole or vertex. For if any other straight line has been drawn through B, e.g., PR, then it forms an angle with the active line AC, e.g., RBC; if then the straight line BH is drawn from the pole to the directrix in such a way that the angle DBH equals the angle RBC and if a line is drawn through the point H parallel to the diameter, let us say H, then this straight line will be the generator and HBR the moving angle, for it equals the moving angle DBC; BR, however, is the active arm and so the point where HG and BR meet lies on the parabola. Because the straight line PR intersects the parabola not only at the point B, but also at the point G, and because the whole straight line BG[3] lies inside the curve, therefore the straight line will not touch the parabola but will intersect it. Therefore all chords drawn in a parabola parallel to the tangent at the vertex are ordinate-wise applied to the diameter, i.e., they are bisected by this diameter; conversely any straight line that is drawn through the vertex, parallel to a chord, i.e., bisected by the diameter, will touch the parabola at the vertex.

Corollary 10

From the preceding it is also clear how to describe a parabola in the plane when given the position of the diameter, its vertex and the latus rectum as well as the angle included by this diameter and the chords that are ordinate-wise applied to it.

For let BK be a diameter of the parabola to be described, B its top, BD the latus rectum belonging to this diameter (measured on this diameter in the appropriate direction) and ABK or CBK the angle between the said diameter and the chords that are ordinate-wise applied to it.

[3] Cor. 1, p. [164].

re. Sumpto enim in curva præter *Polum* B puncto utcunque, veluti G, si *crus efficiens* eidem applicetur, uti in positione B G, constituetur ab ipso & *efficiente* angulus, ut G B C: atque adeò punctum G, utcunque sumptum, id est, tota Parabola, præter *Polum* B, infra *efficientem* A B C cadet.

Corollarium 9.

Constat quoque ex antedictis, non posse aliam rectam præter *efficientem* Parabolam in *Polo* seu vertice contingere. Quoniam enim alia quævis recta per B ducta, ex. gr., P R, angulum constituit cum *efficiente* A C, ut R B C, si à *Polo* ad *directricem* ducatur recta B H, ita ut eidem angulo R B C æqualis sit angulus D B H, ac per punctum H agatur recta diametro parallela, ut H G: erit ea ipsa *describens*, & H B R *angulus mobilis*, utpote æqualis *angulo mobili* D B C; B R verò *crus efficiens*: ac proinde ipsarum H G, B R intersectio G in Parabola. Quare cum recta P R non in puncto B solummodo, sed & in puncto G Parabolæ occurrat, ac tota B G recta [3] intra curvam cadat, non continget recta P R Parabolam, sed eandem secabit.

[3] per Corol. 1 hujus.

Itaque omnes rectæ in Parabola ductæ, quæ contingenti in vertice æquidistant, ordinatim ad diametrum applicantur sive ab eadem diametro bifariam dividuntur; & contra, quæ cuilibet rectæ, à diametro bifariam divisæ, per verticem æquidistans ducitur, Parabolam in vertice contingit.

Corollarium 10.

Ex dictis quoque obvium est, quo pacto datâ positione Parabolæ diametro, ejusque vertice, & latere recto, nec non angulo, quem ordinatim applicatæ faciunt ad eandem diametrum, ipsa Parabola in plano describatur. Si enim describendæ Parabolæ diameter sit B K, vertex B, latus rectum ad eandem diametrum pertinens B D, (quod quidem ipsi diametro in directum sit positum,) atque angulus quem faciunt ad dictam diametrum ordinatim applicatæ A B K vel C B K: oportet, ductâ per D. lateris recti termi-

[169] Then we must first draw the straight line EDF through D (the endpoint of the latus rectum), making the angle EDB equal to ABD [1.12] and next we must describe a curve, say NBG, with AC as the active line, BD as its interval and EF as its directrix. This will be the parabola to be described.

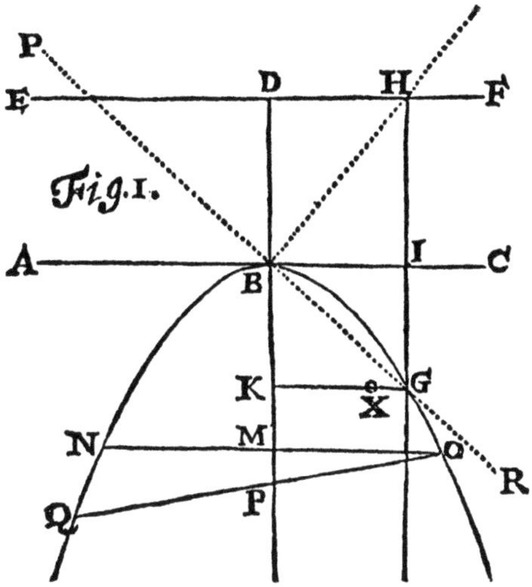

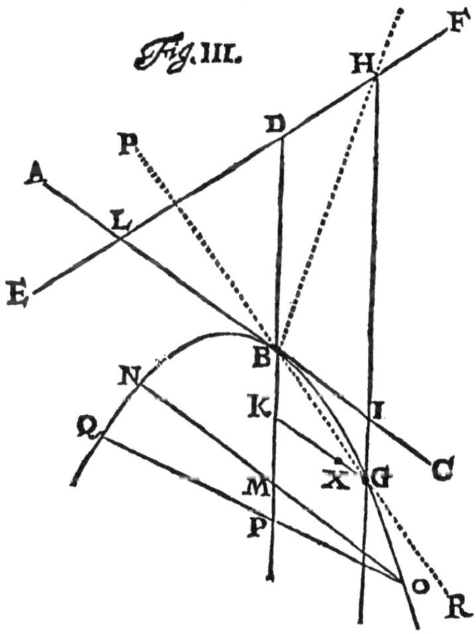

terminum rectâ E D F in angulo E D B ipſi A B D æquali, *efficiente* A C, & *intervallo* B D, ad *directricem* E F curvam deſcribere, ut N B G: eritque hæc ipſa, quæ deſcribenda proponitur Parabola.

[170]

Theorem II

Proposition 2

Let a straight line be drawn through an arbitrary point on a parabola, parallel to the axis or the diameter; then this arbitrarily chosen point will also be a vertex of the parabola and this parallel line will also be a diameter.

Let HAM be an arbitrary parabola with axis or diameter AB and corresponding latus rectum AC, and through an arbitrary point M [1.13] on the curve let there be drawn a straight line MO, parallel to the axis or diameter AB. I say that that the chosen M is a vertex too and that MO is a diameter. Indeed, let a straight line SV be drawn through M, cutting off from the axis or diameter AB a segment AI, outside the parabola, equal to AB (cut off between the vertex A and the ordinate-wise applied BM). Let also OM be produced to K, where K has been chosen in such a way that MK is the third proportional belonging to AB (i.e., AI) and IM. If a parabola is described with SV as the active line and MK as the interval, then I say that this parabola is the same as the given one in this way: that both coincide totally; and I say that not only MO is a diameter and M a vertex, but MK is the latus rectum belonging to the diameter MO, SV touches the parabola at the vertex M, and all line segments parallel to it are bisected by MO and are ordinate-wise applied to MO itself.

Now let another arbitrary point, e.g., H, be chosen on the given parabola HAM, and let from this point be drawn HG, ordinate-wise applied to the axis or diameter AB, as well as HO parallel to SV, the former of which (produced if necessary) meets the straight line KO at E; the latter, however, (produced too if necessary) intersects the said axis or the diameter AB at D. It is also clear that[1] the parabola described by the active line SV and with the interval MK will pass through the point H too, if the square on HO equals the rectangle KMO. The fact that the square on the line segment HO equals the rectangle KMO can be proved in various ways, but, at least in my opinion, simply and rather briefly as follows. Since[2] MB is to BA as CA is to MB,

[1]Prop. 1, p. [162]; [2]Prop. 1, p. [162] and VI, 17.

THEOREMA II.

Propositio 2.

Si per aſſumptum utcunque in Parabola punctum recta ducatur, axi diametrovè parallela, erit quoque aſſumptum punctum Parabolæ vertex, ductaque parallela itidem diameter.

Sit Parabola quælibet H A M, cujus axis diametervè A B, & latus rectum ad eandem pertinens A C; ſitque per punctum M, in curva utcunque aſſumptum, ducta recta M O, axi ſive diametro A B parallela: dico aſſumptum quoque punctum M verticem, dictamque M O diametrum eſſe; imò ſi ductâ, per M rectâ S V, ita ut ab axe ſive diametro A B extra Parabolam abſcindat portionem A I æqualem A B, quæ inter verticem A & applicatam M B intercipitur; productâque O M ad K, ita ut ſit M K ipſis A B vel A I & I M tertia proportionalis, *efficiente* S V, *intervallo* verò M K Parabola deſcribatur: dico hanc cum expoſita Parabola H A M eandem fore, ita ut altera alteri per omnia congruat, ac proinde non ſolùm M O diametrum, atque M verticem fore, ſed & M K latus rectum eſſe ad dictam diametrum M O pertinens, & S V Parabolam in vertice M contingere, omnesque ipſi parallelas in Parabola ductas ab M O bifariam dividi, atque ad hanc ipſam M O ordinatim applicari.

Sit enim in expoſita Parabola H A M aſſumptum præterea aliud quodpiam punctum, ex. gr., H; ſitque ab eodem ducta H G ad axem ſive diametrum A B ordinatim applicata, nec non H O ipſi S V æquidiſtans, quarum prior, ſi opùs fuerit producta, rectæ K O occurrat in E; poſterior verò, itidem producta, ubi opùs fuerit, prædictum axem ſive diametrum A B ſecet in D. Et apparet [1], ſi quadratum rectæ H O æquale ſit rectangulo K M O, Parabolam, quæ *efficiente* S V, *intervallo* verò M K deſcribetur, per punctum quoque H tranſituram. Eſſe autem quadratum rectæ H O æquale rectangulo K M O multifariam id quidem, &, meo ſaltem judicio, breviter ſimpliciterque ſatis in eum qui ſequitur modum demonſtratur.

Quoniam eſt [2] ut C A ad M B, ita M B ad B A, erit, duplicatis

[1] ex 1 *hujus.*

[2] per 1 *hujus,* & 17 *ſexti.*

[171]

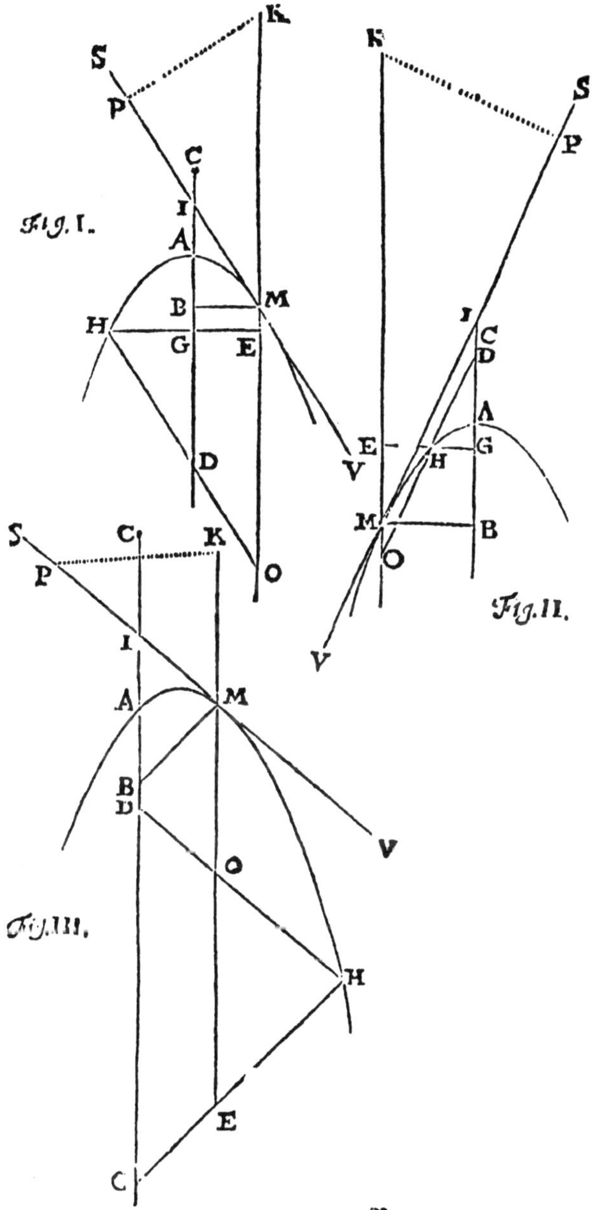
K
S
P
C
I
Fig. I.
A
M
B
H
G
E
D
V
O
K
S
P
I
C
D
A
E
G
H
M
B
O
Fig. II.
V
S
C
K
P
I
A
M
B
D
V
O
Fig. III.
H
E
C

[172] therefore is–by doubling the consequents–MB is to BI (i.e.,[3] HG is to GD) as CA is to the double of MB or twice GE. Therefore[4] the area contained by the means, i.e., twice the rectangle HGE, equals the area contained by the extremes, namely the rectangle enclosed by CA and GD. Both squares on the line segments HG and BM or GE are equal[5] to both rectangles CAG and CAB or CAI, i.e.,[6] the rectangle contained by CA and IG; therefore–if we add or subtract on both sides the same magnitudes, namely twice the rectangle HGE on one side and the rectangle contained by CA and GD on the other side[7]–then the sum[a] or the remainder[b], namely the square on EH and the rectangle contained by CA and ID or MO, will be equal.

The square on EH is to the square on HO[11] as the square on BM is to the square on MI, or as the rectangle CAB[8] is to the rectangle contained by KM and AB,[9] i.e.,[10] as CA is to KM, or–if we assume a common altitude MO –as the aforementioned rectangle contained by CA and MO is to the rectangle KMO. Moreover, since –as shown before–the rectangle contained by CA and MO equals the square on EH, therefore the rectangle KMO will also be equal to the square on HO.[12] As the point H–wherever chosen on the given parabola AH–always also lies on the parabola described by the active line SV, but with the interval MK, so it follows that both parabolas coincide entirely and one is the same as the other. So the truth of the proposition has been established.

Corollary 1

From the foregoing the following is clear: if one draws in a parabola two arbitrary line segments, parallel to each other, then the line that bisects both of them will prove to be a diameter of this parabola. Indeed, the diameter drawn through the midpoint of one of them–be it the diameter by which the parabola has been generated or a diameter parallel to that–

[3]I, 29 and VI, 4; [4]VI, 16; [5]Prop. 1, p. [162]; [6]II, 1; [a]in the case of Fig. I and the like [b]in the case of Figs. II and III and the like; [7]according to what has been proved above, twice the rectangle HGE is equal to the rectangle contained by CA and GD; [c]in the case of Fig. I, if on one side the rectangle HGE is added twice to the squares on HG and GE, the result is –according to II, 4–the square on EH as sum; if on the other side the rectangle contained by CA and GD is added to the rectangle contained by CA and IG, the result is–according to II, 1–the rectangle contained by CA and ID (or MO). In a similar way we see in Figures II and III: if on one side the rectangle HGE is subtracted twice from both squares HG and GE, the residue will be the square on EH, according to II, 7, and if on the other side the rectangle contained by CA and GD is subtracted from the rectangle contained by CA and IG, the residue will be the rectangle contained by CA and ID, or MO, according to II, 7. [8]Prop. 1, p. [162] and by hypothesis. [9]VI, 17 and by hypothesis; [10]VI, 1; [11]VI, 4 and 22; [12]V, 14.

catis conſequentibus, ut C A ad duplam M B ſeu ad G E bis, ita M B ad B I, hoc eſt [3], ita H G ad G D: ac proinde [4] contentum ſub mediis, nempe rectangulum H G E bis, æquale contento ſub extremis, nimirum rectangulo ſub C A & G D. Vnde cum bina quadrata rectarum H G & B M ſeu G E æqualia ſint [5] binis rectangulis C A G & C A B ſeu C A I, id eſt [6], rectangulo ſub C A & I G: erunt quoque, additis [a] demptiſve [b] utrinque æqualibus, nimirum rectangulo H G E bis ab una, ac rectangulo ſub C A & G D ab altera parte [7], compoſita [a] vel reſidua [b], nempe quadratum E H ac rectangulum ſub C A & I D ſeu M O æqualia [c].

[3] per 29 primi, & 4 ſexti. [4] per 16 ſexti. [5] per 1 hujus. [6] per 1 ſecundi. [a] in caſu fig. I & ſimilibus. [b] in caſu fig. II & III ac ſimilibus. [7] *quippe per ſupra demonſtrata rectangulum* H G E *bis æquale eſt rectangulo ſub* C A *&* G D. [c] in caſu enim fig. I, ſi ab una parte ad bina quadrata rectarum H G & G E addatur rectangulum H G E bis, compoſitum fit E H quadratum, per 4 ſecundi; ac ſi ab altera parte ad rectangulum ſub C A & I G addatur rectangulum ſub C A & G D, ſit, per I ſecundi, rectangulum ſub C A & I D ſeu M O. Eodem modo, ſi in caſibus fig. II & III ab una parte à binis quadratis rectarum H G & G E auferatur rectangulum H G E bis, reſiduum erit, per 7 ſecundi, E H quadratum; ac ſi ab altera parte à rectangulo ſub C A & I G auferatur rectangulum ſub C A & G D reſiduum erit, per 1 ſecundi, rectangulum ſub C A & I D ſeu M O.

Ideoque cum ſit ut B M quadratum ad M I quadratum, ſive ut C A B rectangulum [8] ad rectangulum ſub K M & A B [9], hoc eſt [10], ut C A ad K M, ſeu, aſſumptâ communi altitudine M O, ut prædictum rectangulum ſub C A & M O ad K M O rectangulum, ita [11] E H quadratum ad H O quadratum; ſitque rectangulum ſub C A & M O, ut jam oſtenſum eſt, æquale quadrato E H: erit quoque [12] rectangulum K M O quadrato H O æquale.

Vnde cum punctum H, ubicunque id in expoſita Parabola A H aſſumptum fuerit, ſemper quoque ſit in Parabola, quæ *efficiente* S V, *intervallo* verò M K deſcribitur: ſequitur alteram alteri per omnia congruere, ideoque hanc cum illa eandem eſſe; ita ut conſtet veritas eorum, quæ proponebantur.

[8] per 1 hujus, & ex hypotheſi. [9] per 17 ſexti, & ex hypotheſi. [10] per 1 ſexti. [11] per 4 & 22 ſexti. [12] per 14 quinti.

Corollarium I.

Ex antedictis manifeſtum eſt, quòd, ductis in Parabola binis quibuſlibet rectis ſibi invicem æquidiſtantibus, quæ utramque bifariam dividit recta linea illius diameter exiſtat. Quippe quæ per medium æquidiſtantium unius diameter ducetur, ſive hæc ſit ipſa diameter ex generatione, ſive eidem parallela, per medium quo-

[173]

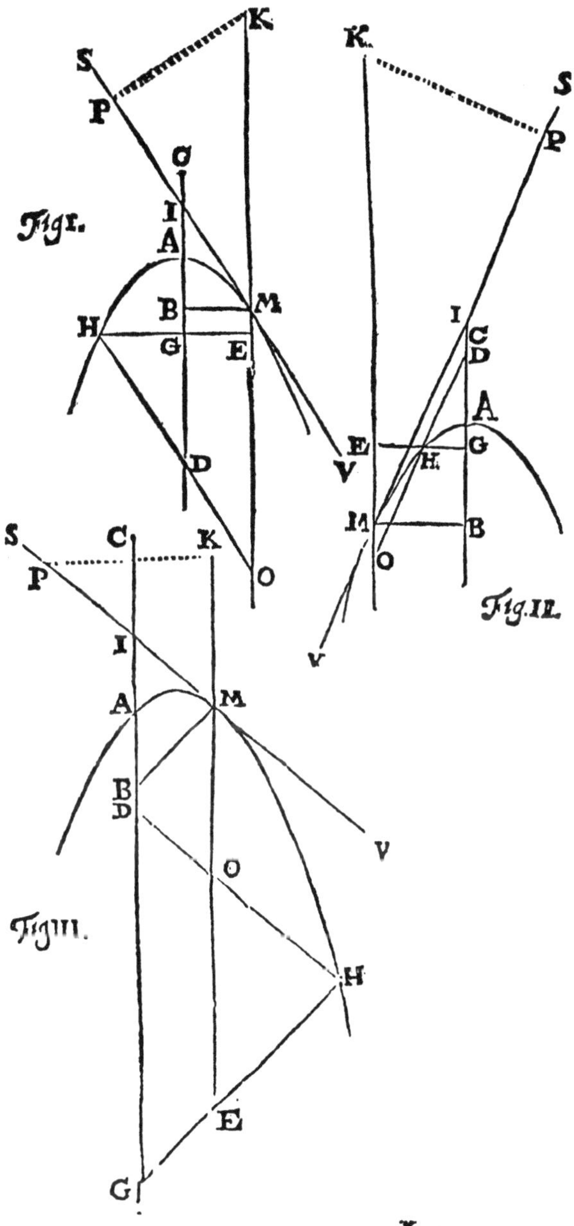
Fig. I.
S
P
K
C
I
A
B
M
H
G
E
D
V
O
Fig. II.
K
S
P
I
C
D
A
E
G
H
M
B
O
V
Fig. III.
S
C
K
P
I
A
M
B
D
V
O
H
E
G

[174] also passes through the midpoint of the other of the two parallel line segments.[1] And so it is clear how to find a diameter of an arbitrary parabola and, at the same time, the lines that are ordinate-wise applied to it.

Corollary 2

Furthermore, it is clear that arbitrary straight lines touching a parabola, wherever it may be, and lines that are ordinate-wise applied to the diameter from the point of contact, cut off equal parts of the diameter on both sides of the vertex and vice versa; if one draws from the endpoints (on the parabola, transl.) of lines that are ordinate-wise applied to a diameter, other lines such that these lines and the ordinate-wise applied lines cut off equal parts of the diameter on both sides of the vertex, then these lines will touch the parabola at the said endpoints. Indeed, we just[2] proved that the straight line SV touches the parabola at the arbitrarily chosen point M because AI and AB are equal; however, we proved above[3] that no other line can touch the parabola at M.

Corollary 3

From this we can easily understand how to draw from an arbitrary point, not within a given parabola, a straight line that touches the parabola [1.17]. Indeed, if we have found[4] some diameter and the lines that are ordinate-wise applied to it and if the endpoint of this diameter coincides with the given point, then it is now known[5] that the straight line drawn through this point parallel to the ordinate-wise applied lines will touch the parabola at this point. But if elsewhere on the curve a point has been given, e.g., M, and if a diameter AD has been found, then we must first draw from M the straight line MB, ordinate-wise applied to AD, next take AI equal to AB, and then draw the straight line through I and M. If, however, a point is given outside the curve, on the produced part of a diameter, e.g., I, then we must first make AB equal to AI and apply BM ordinate-wise to AD; let this line (BM, transl.) intersect the parabola at M, then we must next draw a straight line through I and M. But if the point is given neither on the curve nor on the produced part of a diameter, e.g., if we have found a diameter MO and if the given point is I, then we must first draw ID, parallel to the diameter MO, which intersects the parabola at A, then we take AB equal to AI, and draw from B the line BM ordinate-wise applied to AD, i.e., a line parallel to the tangent at A, intersecting the parabola at M, and finally we draw a straight line through I and M, for it has been established above[6] that IM touches the parabola at the point M anyhow.

---

[1]Prop. 1, Cor. 6, p. [166]; [2]Prop. 2, p. [170]; [3]Prop. 1, Cor. 9, p. [168]; [4]Prop. 2, Cor. 1, p. [164]; [5]Prop. 1, Cor. 8 and 9, p. [167], [168]; [6] Prop. 2, p. [170] and Cor. 2, p. [174].

quoque alterius æquidiſtantium tranſibit [1]. Atque ita apparet, quo pacto datæ cujuſlibet Parabolæ diametrum ſimulque ordinatim ad eandem applicatas invenire liceat.

[1] per conclusionem 6 Cor. 1 hujus.

Corollarium 2.

Patetque porrò, quaſlibet rectas Parabolam ubivis contingentes, atque ordinatim à puncto contactus ad diametrum applicatas, æquales utrinque à vertice diametri portiones abſcindere; &, vice versâ à terminis applicatarum per diametrum ductas, ita ut æquales utrinque à vertice diametri portiones ductæ applicatæque abſcindant, Parabolam in dictis terminis contingere. Rectam enim S V, ex eo quòd æquales ſint A I, A B, Parabolam in puncto M utcunque aſſumpto contingere, nunc [2] demonſtratum; at nec aliam rectam in puncto M Parabolam contingere poſſe, ſuperiùs [3] oſtenſum eſt.

[2] in 2 hujus.
[3] per 9 Cor. 1 hujus.

Corollarium 3.

Atque hinc non difficulter colligitur, quo pacto à quolibet puncto, non intra Parabolam dato, recta ducatur, quæ Parabolam contingat. Inventis enim [4] diametro quâcunque & rectis, quæ ad illam ordinatim applicantur, ſi in ejuſdem diametri termino ſit datum punctum, notum nunc eſt [5] rectam per idem punctum ductam, atque ordinatim applicatis æquidiſtantem, Parabolam ibidem contingere. At ſi alibi in curva ſit punctum datum, veluti M, ſitque inventa diameter A D: oportet, ductâ ex M rectâ M B ipſi A D applicatâ, ſumptâque A I ipſi A B æquali, ducere rectam per I & M. Sin autem extra curvam detur in diametro producta, veluti I: oportet, factâ A B ipſi A I æquali, atque B M ordinatim ad A D applicatâ, quæ Parabolæ occurrat in M, ducere rurſus rectam per I & M. At verò ſi neque in curva neque in diametro producta detur, ut, ſi inventa diameter ſit M O, datumque punctum I: oportet, ductâ I D diametro M O parallelâ, quæ Parabolam ſecet in A, ſumptâque A B ipſi A I æquali, atque ex B ductâ B M ordinatim ad A D applicatâ, nimirum, quæ æquidiſtans ſit contingenti in A, Parabolæque occurrat in M, ducere iterum rectam per I & M. quippe conſtat ex antedictis [6], ipſam I M omni caſu Parabolam contingere in puncto M.

[4] per 1 Corol. 2 hujus.
[5] per 8 & 9 Corol. 1 hujus.
[6] per 2 hujus, ejuſque Cor. 2.

Co-

[175]

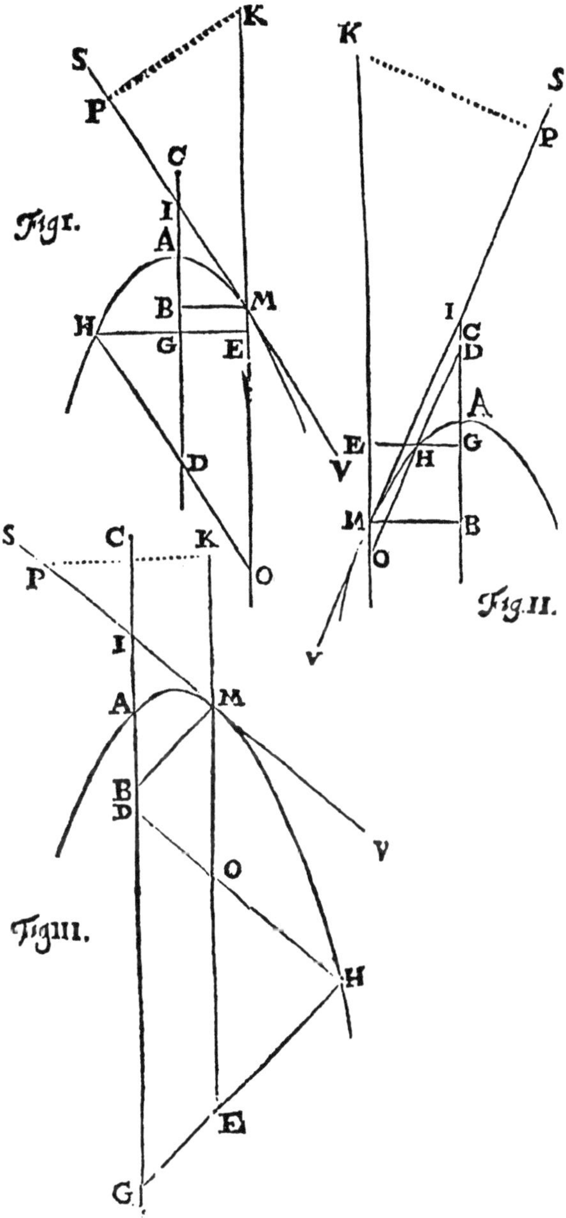
Fig. I.
S
P
K
C
I
A
B
M
H
G
E
D
V
O
Fig. II.
K
S
P
I
C
D
A
E
G
H
M
B
O
V
Fig. III.
S
C
K
P
I
A
M
B
D
O
V
H
E
G

[176] Corollary 4

Moreover, it is a fact [1.18] that the parameter belonging to an arbitrary diameter is the third proportional to two line segments, one of which is that part of the axis or of the given diameter cut off between the vertex and the line that touches the parabola at the endpoint of the chosen diameter; the other, however, is that part of the said tangent that lies between the given and the chosen diameter. Indeed it has been proved[1] that the line segment MK is the parameter belonging to the aribitrarily chosen diameter MO because it is the third proportional to AI and IM.

Corollary 5

Let the position of an arbitrary diameter of a parabola, its vertex and its latus rectum as well as the angle enclosed by this diameter and the lines that are ordinate-wise applied to the said diameter all be given. From what has been proved before, it is easy to understand how to find another diameter of the same parabola–in such a way that it encloses another arbitrarily given angle with lines that are ordinate-wise applied to it–as well as its vertex and its latus rectum. Indeed, let the position of the diameter MO, its vertex M, its latus rectum MK, the angle SMK or VMK enclosed by the said diameter and the lines applied to it all be given and let it be required to find another diameter of the same parabola with which the applied lines form an angle equal to an arbitrarily given angle ABM. Then we must draw from the endpoint K a straight line KP to SV at an angle KPV equal to the given ABM; next we have to bisect PM at I and to draw through I a straight line IB parallel to MO. Thereafter we must draw the line segment MB from M to IB at an angle MBI equal to the given angle. When BI has been bisected at A, then the desired diameter will be AB, its vertex the point A and its parameter AC, namely the line segment that appears as the third proportional to AB and BM. Indeed,[2] the point M lies on the parabola that is described by an active line parallel to BM itself and with an interval AC, as by construction[3] the square on the applied BM equals the rectangle CAB. As the triangles BIM and PMK are similar–because the angles at B and P are equal (by construction) to those at I and M[4] (because AD and MO are parallel)– therefore PM will be to MK as BI is to IM[5] and–by bisecting the antecedents– IM is to MK as AI is to IM. Because of what has been proved before,[6] the parabolas with diameters AB and MO and parameters AC and MK

[1]Prop. 2, p. [170]; [2]Prop. 1, p. [162]; [3] VI, 17; [4]I, 29; [5]VI, 4; [6]Prop. 2, p. [170].

Corollarium 4.

Conſtat præterea, aſſumptæ cujuſlibet diametri parametrum eſſe tertiam proportionalem duabus rectis, quarum una eſt vel axis vel datæ diametri portio, intercepta inter ejuſdem verticem & eam, quæ Parabolam in aſſumptæ diametri termino contingit, altera verò ea prædictæ contingentis pars, quæ inter datam & aſſumptam diametrum interjacet. Demonſtratum enim eſt [1], rectam MK, ex eo quòd ipſis AI, IM tertia ſit proportionalis, aſſumptæ utcunque diametri MO parametrum eſſe.

[1] *in 2 hujus.*

Corollarium 5.

Ex demonſtratis quoque non difficulter colligitur, quo pacto, datâ poſitione quâlibet Parabolæ diametro, ejusque vertice, & latere recto, nec non angulo, quem faciunt ordinatim ad dictam diametrum applicatæ, alia ejuſdem Parabolæ diameter, quâcum applicatæ alium quemlibet angulum conſtituant, ac ipſius vertex, & latus rectum inveniantur. Si enim datâ poſitione diametro MO, vertice M, & latere recto MK, anguloque SMK vel VMK, quem applicatæ faciunt ad dictam diametrum MO, aliam ejuſdem Parabolæ diametrum invenire oporteat, quâcum applicatæ angulum conſtituant æqualem dato cuilibet angulo ABM: ducatur à termino K ad SV recta KP in angulo KPV ipſi dato ABM æquali, divisâque PM bifariam in I ducatur per I recta IB ipſi MO æquidiſtans. Deinde ab M ad eandem IB applicetur recta MB in angulo MBI dato angulo æquali, divisâque BI bifariam in A, erit quæſita diameter AB, vertex punctum A, ejusque parameter AC, recta nempe, quæ ipſis AB, BM tertia proportionalis exiſtit. Eſt enim [2] punctum M in Parabola, quæ *efficiente* ipſi BM parallelâ ac *intervallo* AC deſcribitur, quandoquidem quadratum applicatæ BM ex conſtructione [3] rectangulo CAB eſt æquale. Deinde quoniam ſimilia ſunt triangula BIM & PMK, ob æquales angulos ad B & P (ex conſtructione), atque ad I & M [4] (ob parallelas AD, MO) erit [5] ut BI ad IM, ita PM ad MK. &, ſumptis antecedentium dimidiis, ut AI ad IM, ita IM ad MK. Quare ſecundùm ea quæ ſuperiùs [6] demonſtrata ſunt, Parabolæ diametris AB, MO, ac parametris AC, MK,

[2] *per 1 hujus.*
[3] *per 17 ſexti.*
[4] *per 29 primi.*
[5] *per 4 ſexti.*
[6] *in 2 hujus.*

[177]

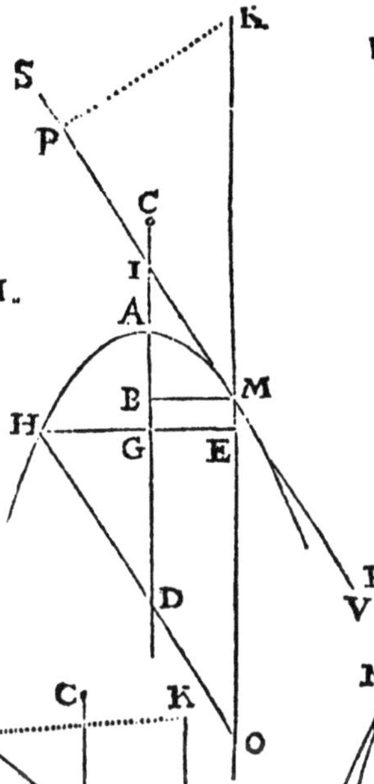

Fig. I.

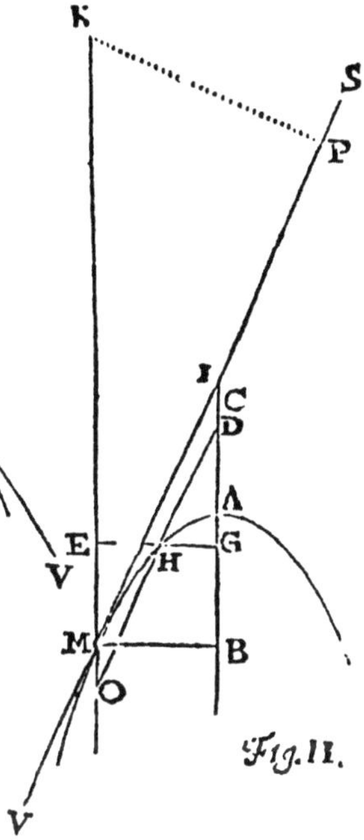

Fig. II.

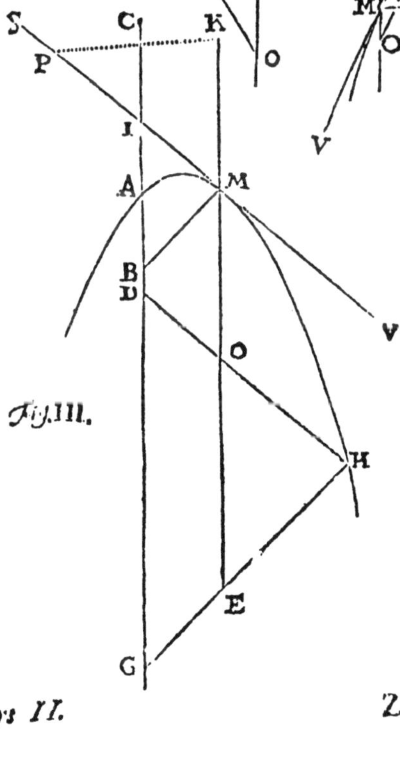

Fig. III.

[178] and described with the said angles [1.19], are exactly the same. However, the angles formed by MB and the other straight lines that are ordinatewise aplied to the diameter AD, are also equal to the given angle ABM by construction. So what was asked has been achieved. If, however, the given angle should be a right angle, then the AD we found will be the axis.

The curve generated by an arbitrary active line and an arbitrary interval is –if the moving angles differ from the angles at the directrix on the same side (of the interval, transl.)–the one to which I allot the first place among the curves of the first genus [1.20] after the circle and the parabola, as I think it is in a sense simpler than the following species. Therefore in explaining its origin, its nature, and its properties–which I am going to do–I could pursue the same method of describing and adhere to the definitions I gave at the beginning [1.21]. One of its properties, however, which in my opinion is the most important and the most general and from which the other properties can be derived very easily, comes up more clearly and is more readily shown, which in mathematics and especially when explaining the foundations, is in my opinion very important. Therefore I chose the property that deviates as little as possible from the method mentioned before. This also operates by the movement and intersection of a rectilinear angle and a straight line, but now the movement of the said angle is not circular, but rectilinear, and on the other hand the movement of the said straight line is not rectilinear, but circular, as will be more clear from the definitions that have been adapted to this purpose and are set forth in the next chapter.

Chapter II

Second Definitions

I

When a straight line rotates about a fixed point and at the same time pushes a rectilinear angle along a fixed straight line so that one of its arms lies on

MK, in dictis angulis descriptæ omnino eædem erunt. Sunt autem & anguli, quos faciunt MB aliæque ad diametrum AD applicatæ, ex constructione, dato angulo ABM æquales. Quocirca effectum est, quod quærebatur. Quòd si verò datus angulus ABM rectus fuerit, ipse axis erit, inventa AD.

Etiamsi curva, quâlibet *efficiente*, & quocunque *intervallo* descripta, si *anguli mobiles* inæquales sint iis, qui ad *directricem* sunt ab eadem parte, ea ipsa sit, cui post Circulum & Parabolam inter curvas primi generis primum locum tribuam, utpote quam sequenti specie quodammodo simpliciorem judicem; cujusque propterea ortum, naturam, & proprietates nunc expositurus eidem describendi methodo insistere, præmissisque in principio definitionibus inhærere possem; cum tamen ea ipsius proprietas, quam primam ac maximè universalem existimo, & è qua cæteras facillimè deduco, ex aliis generationum speciebus distinctiùs appareat atque expeditiùs demonstretur, quod in Mathematicis, & præcipuè in Elementorum explicatione non parvi faciendum puto, eam selegi, quæ à jam dicta quàm minimum deflectat, quæque similiter anguli rectilinei rectæque lineæ motu & intersectione perficitur; at in qua dicti anguli motus non circularis sed rectus, ac contra dictæ lineæ non rectus sed circularis est, ut ex definitionibus in eum finem adaptatis, & sequenti Capite propositis, magis elucescet.

CAPUT II.

DEFINITIONES SECUNDÆ.

I.

SI recta linea circa punctum fixum circulariter mota angulum quendam rectilineum, altero sui crure immo-

[179] this fixed straight line, and so that the rotating straight line always passes through the same point of the (on the fixed line) applied arm, and so that at the same time the point of intersection of the other arm and the rotating line describes a curve, then this rotating straight line will be called the generator [2.1].

II

The other, fixed straight line will be called the directrix.

III

The said angle and its adjacent angle will bear the name moving angles–as in the former case.

IV

Likewise, the fixed point about which the generator rotates will be called the pole.

V

The arm of the moving angle on the other hand that is pushed along the directrix by the generator will be called the passive arm.

VI

The other arm, which is intersected by the generator, will be called the active arm; when it is produced through the vertex of the moving angle, it will be called the active line.

VII

When the generator is parallel to the active line and so they do not intersect, we say that both the active line and the generator are in the initial position, and when they are referred to in this way, we assume that they are in this position.

immotæ rectæ lineæ applicatum, per eandem immotam lineam promoveat, & secum ducat, ita ut prædicta recta circulariter mota semper per idem applicati cruris punctum transeat, simulque alterius cruris ac ejusdem lineæ motæ intersectione curva describatur, appellabitur hæc ipsa circulariter mota *linea describens*.

II.

Altera verò immota manens *Directricis* nomen retinebit.

III.

Prædictus autem angulus rectilineus, isque qui ei est deinceps, similiter & hîc *Angulorum mobilium* nomine venient.

IV.

Sicuti & punctum fixum, circa quod *describens* circulariter movetur, *Polus* nuncupabitur.

V.

Rursusque crus *anguli mobilis*, quòd à *describente* per *directricem* promovetur, *Crus patiens*.

VI.

Alterum autem crus, quod à describente secatur, *Crus efficiens*, & per anguli verticem productum *Linea efficiens* appellabitur.

VII.

Cùm *describens efficienti* parallela est ac proinde nulla ipsarum intersectio existit, tam *efficientem* quàm *describentem* in *statione prima* constitutas dicemus; ac quoties de iis simpliciter sermo erit, in tali ipsas statione considerabimus.

[180]

VIII

Here, however, we give the name interval to that part of the passive arm that lies between the vertex of the moving angle and the generator, as well as to that part of the generator (in the initial position, transl.) that is cut off between the pole and the directrix. We refer to the added figures. If we imagine that the line *ABC*[a] rotates about the point *A* and that during its motion it pushes forward and carries along the angle *BEC*,[a] so that the arm *EB* always remains attached to the fixed straight line *KL* and so that the said moving *ABC* always passes through the same point on the arm *EB*, let us say through the point *B*, and so that at the same time the point of intersection of the other arm–*EC*–with the said *ABC* describes a curve *cC*, while the drawn straight line *AD* is parallel to *EC*, then it is clear that the closer the straight line *ABC* approaches *AD*, the smaller the angle *ECB* will be and that in the end, when *ABC* reaches *AD* so that *ABC* coincides with *AD*, this angle *ECB* completely vanishes. The reason is that *AD* and therefore also the said *ABC* in this position are parallel to the arm *EC*. In this situation the said arm *EC* or the straight line *CEM* is the same as the straight line *GFH*: for indeed we supposed that *DF* is equal to *BE*, and so the following will be true:

ABC is the generator in different positions.
KL is the directrix.
BEC, *BEM*, or *DFH* and *DFG* are the moving angles.
A is the pole.
EB is the passive arm; *EC* is the active arm.
MC is the active line; *GFH* is the active line in the initial position or simply the active line.
ADI is the generator in the initial position or simply the generator.
EB–or *FD*–and *AD* are both the interval.

Theorem III

Proposition 3

A curve described according to the preceding definitions, by arbitrary moving angles and arbitrary intervals, has the property that the rectangle

[a]This straight line *ABC* is, like the angle *BEC*, displayed in the figure in four distinct positions.

VIII.

Intervallum autem hic nominabimus tam eam *Cruris patientis* partem, quæ inter *anguli mobilis* verticem & *describentem* interjacet, quàm eam *describentis* portionem, quæ inter *Polum* & *directricem* intercipitur.

Vt in apposita figura, si recta A B C [a] circa A punctum circulariter moveri concipiatur, motuque suo promovere & secum ducere angulum B E C [a]; ita ut crus E B semper applicatum maneat immotæ rectæ lineæ K L, ac prædicta A B C mobilis semper transeat per idem punctum cruris E B, ex. gr., per B, simulque alterius cruris E C & dictæ lineæ A B C intersectione C describatur curva linea *c* C, sitque ducta A D cruri E C parallela: apparet, quò magis recta A B C ad ipsam A D accedit, eò minorem fieri angulum E C B. ac tandem cùm ipsa A B C pervenit ad A D, ita ut cum ipsa coincidat, eundem angulum E C B tunc penitus evanescere: cum A D, ac proinde & dicta A B C, statione illâ, cruri E C parallela sit; ita ut tunc dictum crus E C sive recta C E M eadem sit cum linea G F H, nimirum suppositâ D F ipsi B E, æquali, eruntque

[a] quæ quidem rectæ A B C, ut & angulus B E C in figura quatuor distinctis stationibus exhibentur.

A B C *describens* in stationibus diversis.
K L *directrix*.
B E C, B E M, sive D F H & D F G *anguli mobiles*.
A *Polus*.
E B *crus patiens*.
E C *crus efficiens*.
M C *linea efficiens*.
G F H *efficiens in statione prima*, seu *efficiens* simpliciter.
A D I *describens in statione prima*, seu *describens* simpliciter.
E B seu F D & A D utrumque *intervallum*.

THEOREMA III.

Propositio 3.

Quibuslibet *angulis mobilibus* ac quibuscunque *intervallis*, juxta definitiones præmissas descriptâ curvâ, hoc ipsi proprium erit, ut rectangulum contentum

[181] contained by an arbitrary line segment, drawn from any point on the curve, parallel to the active line and by that part of the directrix, cut off between the said line segment and the active line (in the initial position), equals the rectangle contained by both intervals [2.2].

Let the curve cC be described by an arbitrary moving angle BEC and by arbitrary intervals EB–or FD–and AD and by the directrix KDL so that the active line is GFH (in the initial position, transl.) and let be drawn CE, from an arbitrary point C on the curve to the directrix, parallel to the active line GFH and therefore also parallel to the interval AD. I say that the rectangle FEC equals the rectangle ADF or the rectangle contained by AD and EB [2.3].

tum sub qualibet recta *efficienti* parallelâ, à quocunque curvæ puncto ad *directricem* ductâ, atque eâ *directricis* parte, quæ inter dictam parallelam & *efficientem* intercipitur, æquale sit ei, quod sub utroque *intervallo* continetur, rectangulo.

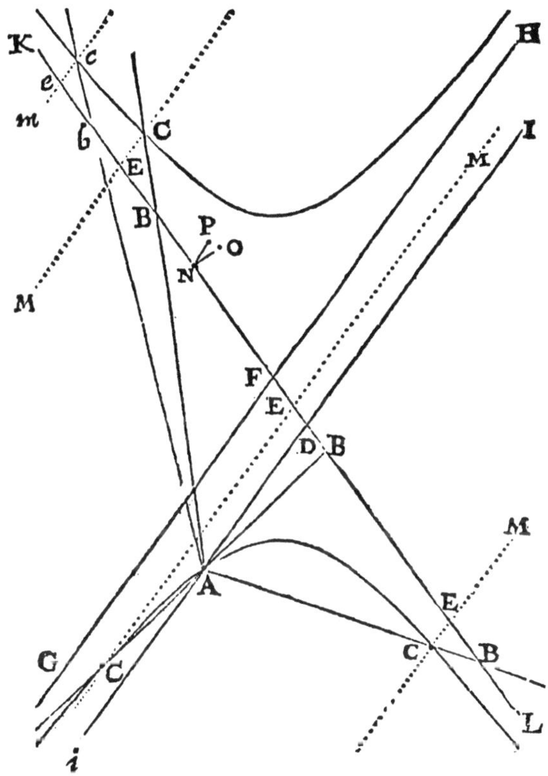

Sit quolibet *angulo mobili* B E C, & quibuscunque *intervallis* E B, seu F D & A D, *directrice* K D L, descripta curva *c* C; ita ut *efficiens* sit G F H, sitque à puncto C in curva utcunque assumpto ad *directricem* ducta C E *efficienti* G F H, ac proinde & *intervallo* A D parallela: dico rectangulum F E C æquale esse A D F rectangulo, sive ei, quod sub A D, E B continetur.

[182] Indeed, if we bring the moving angle as well as the generator into the position in which they were when they generated by their intersection the point C (let us say in the position of BEC and ABC), then the line segments BD and FE will also be equal, because the line segments EB and FD are equal, and we can add or subtract FB or ED on both sides [2.4]. Since[1] EC and AD are parallel, the triangles BDA and BEC are equiangular; therefore[2] BD, i.e., FE, is to DA as BE is to EC and so[3] the rectangle FEC contained by the extremes equals the rectangle contained by the means AD and EB or the rectangle ADF. This was the statement.

Because all rectangles as FEC are also mutually equal [2.5], it is clear that the curve, described by intersections in the preceding way, is the same curve that the Ancients called a hyperbola and that–if we consider the two curves simultaneously generated by the same continuous motion–those are the curves that they called the opposite branches; it is also clear that the directrix KL and the active line (in the initial position) GH, are those that they called the asymptotes and their point of intersection–F–is the same point that they called the center of the hyperbola or the center of the opposite branches. Therefore we will in the future also refer to each of those concepts by the same name. We will, however, avoid the name *sections* by the reasons explained before,[4] because we think these are less appropriate. We will, however, call the rectangle contained by the intervals or the square equal to it, the *power* of the hyperbola.

Corollary 1

From the description itself it is clear that the asymptotes and the hyperbola approach each other continuously more and more and that finally they reach a distance smaller than any given distance. If, however, you desire a more exact proof, consider the following: Let this distance be given as a line segment NO, perpendicular to the asymptote FK. If we choose NP, smaller than NO, and if DF is to Fe as NP is to AD and if through e the line segment ec is drawn, equal to NP and parallel to the asymptote FH, then[5] the rectangle Fec equals the power ADF

[1] I, 29; [2] VI, 4; [3] VI, 16; [4] in my letter to van Schooten; [5] VI, 16.

Constitutis enim tam *angulo mobili* quàm *describente* in statione uti fuêre, cùm per ipsarum intersectionem descriptum est punctum C, veluti in B E C, & A B C, quoniam æquales sunt rectæ E B, F D, additâ vel ablatâ utrinque F B vel E D: erunt quoque rectæ B D, F E æquales; cumque [1] propter parallelas E C, A D æquiangula sint triangula B D A, B E C: erit [2] ut B D, id est, F E, ad D A, ita B E ad E C: ideoque [3] rectangulum F E C sub extremis æquale rectangulo sub mediis A D, E B seu A D F. Quod erát propositum.

[1] *per 29. primi.*
[2] *per 4 sexti.*
[3] *per 16 sexti.*

Quare cum & omnia rectangula, ut F E C, inter se quoque sint æqualia, manifestum est, curvam, intersectione, uti prædictum est, descriptam, eam ipsam esse, quam Veteres Hyperbolam vocarunt, aut, si binas curvas eodem & continuato motu genitas simul consideres, esse eas, quas Oppositas Sectiones dixêre: *directricem* verò K L ac *efficientem* G H eas ipsas, quas Asymptotos nuncupaverunt, atque ipsarum occursum sive intersectionem, ut F, idem illud punctum, quod Hyperbolæ sive Oppositarum Sectionum Centrum ab ipsis appellatum fuit. ideoque & hæc singula iisdem illis nominibus in posterum indigitabimus, solummodo sectionum nomen, ob rationes superiùs [4] expositas, minùs congruum evitaturi. Rectangulum autem sub *intervallis* contentum, seu, quadratum ei æquale, Hyperbolæ Potentiam dicemus.

[4] *in Epistola ad Schotenium.*

Corollarium I.

Ex ipsa descriptione manifestum est, Asymptotos & Hyperbolam magis magisque ad se invicem continuè accedere, tandemque pervenire ad distantiam, datâ quâlibet distantiâ minorem; cujus tamen, si demonstrationem exactiorem desideres, data distantia sit recta N O, ad Asymptoton F K perpendicularis. Sumptâ igitur N P, quæ eâdem N O minor sit, si fiat ut N P ad A D, ita D F ad F *e*, ac per *e* ducatur *e c* ipsi N P æqualis atque Asymptoto F H æquidistans: erit [5] rectangulum F *e c* Potentiæ A D F æqua-

[5] *per 16 sexti.*

[183] and therefore the point C [2.7] lies on the hyperbola according to what[6] has been proved. However ce also equals PN, which means that it is smaller than the given distance NO. Therefore the perpendicular drawn from the point c to the asymptote FK, i.e., the distance from the hyperbola to the said asymptote, is a fortiori smaller than the distance NO.

Corollary 2

So, at the same time, it is clear that all straight lines drawn from an arbitrary point inside the angle opposite to the angle within which the hyperbola lies, and that pass through the center or intersect one of the two asymptotes, finally meet–when produced–this hyperbola

[6]Prop. 3, p. [180].

æquale, ideoque juxta ea, quæ [6] demonstrata sunt, punctum C in 6 in 3 hu-
Hyperbola. Est autem & *c e* ipsi P N æqualis, hoc est, datâ di- jus.

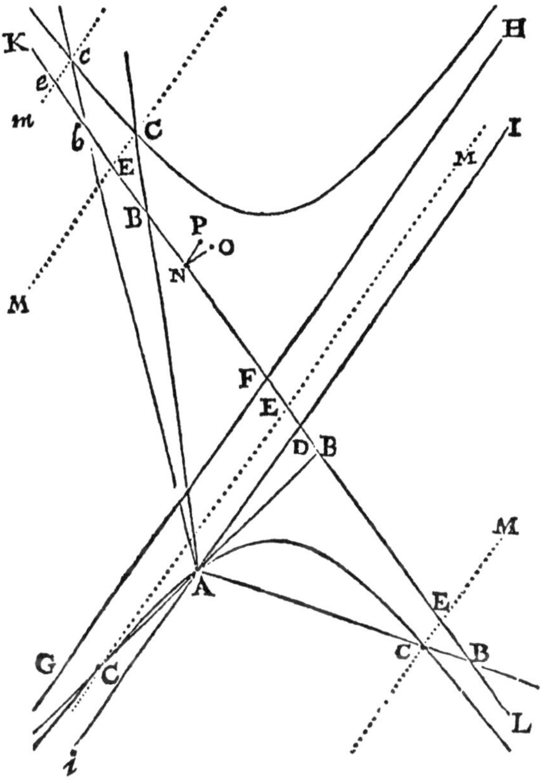

stantiâ N O minor. Quare & perpendicularis à puncto *c* ad Asymptoton F K ducta, id est, distantia Hyperbolæ à prædicta Asymptoto, ibidem datâ distantiâ N O multò minor erit.

Corollarium 2.

Atque ita simul apparet, rectas omnes, quæ ductæ ex quolibet puncto intra angulum, qui ad verticem est ei, qui Hyperbolam continet, per centrum transeunt, vel Asymptotorum alterutram secant, Hyperbolæ tandem occurrere, productasque eandem, & in

[184] and intersect it at only one point, as they -being produced- recede more and more from each of both asymptotes.

Corollary 3

Moreover, it is certain that the active arm in an arbitrary position, i.e., all straight lines parallel to the asymptote [2.8], likewise, meet(s) the hyperbola at only one point, and that they–if produced–intersect the hyperbola there. Indeed it is impossible that the generator and the active arm in any position intersect each other at more than one point.

Theorem IV

Proposition 4

A straight line–whether passing through two arbitrary points on a hyperbola or meeting the hyperbola in such a way that this line, being produced on both sides, lies outside the hyperbola [2.9]–intersects each of the asymptotes within the angle that contains the curve.

Suppose that through the hyperbola BCD, with asymptotes KAE and HAF, the straight line $FBCG$ has been drawn, passing through the two points B and C on the curve, and that MC has also been drawn, meeting the hyperbola in such a way that–being produced towards L–it lies outside the hyperbola on both sides. I say that the straight line $FBCG$ as well as the straight line MCL meet both asymptotes KAE and HAF inside the angle EAF. For, if this should not happen, then $FBCG$ or MCL

184 ELEM. CVRVARVM

in uno tantùm puncto, secare: quandoquidem hæ productæ ab utraque Asymptoto magis magisque semper abscedunt.

Corollarium 3.

Constat præterea, *efficientem* in quacunque statione, id est, rectas omnes Asymptoto parallelas similiter Hyperbolæ, & quidem in uno tantùm puncto, occurrere, productasque illam ibidem secare. Impossibile enim est, ut *describens* atque *efficiens* ullâ statione sese in pluribus punctis intersecent.

THEOREMA IV.

Propositio 4.

Recta linea, sive per bina quælibet in Hyperbola puncta transiens, sive eidem ita occurrens, ut producta utrinque extra Hyperbolam cadat, utrique Asymptoto, intra angulum, qui curvam continet, occurrit.

Sint in Hyperbola BCD, cujus Asymptoti KAE, HAF,

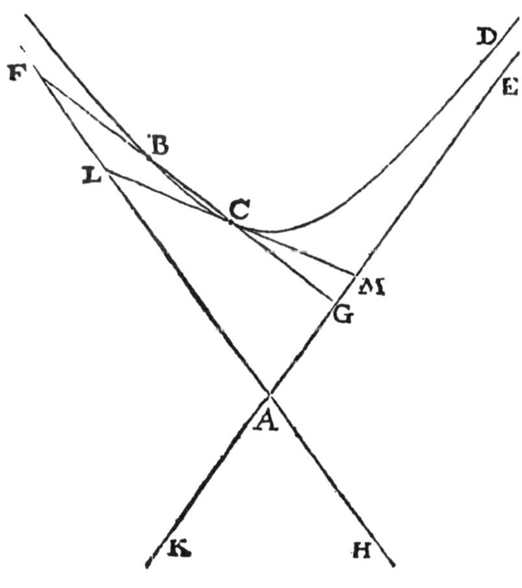

ductæ FBCG, transiens per bina curvæ puncta B & C, atque MC eidem occurrens in C, ita ut producta versùs L utrinque extra Hyperbolam cadat; Dico tam rectam FBCG quàm rectam MCL utrique Asymptoto KAE & HAF intra angulum EAF occurrere. Hoc enim si non accideret, eadem FBCG vel MCL aut Asym-

[185] would either be parallel to one of the asymptotes or–if one of them should meet one of the asymptotes outside the angle EAF–would be a straight line drawn through a point inside the angle KAH, the opposite of the angle that contains the hyperbola, and it would intersect one of the asymptotes. Therefore[1] it would meet the curve at only one point, not at two points, [2.10] and, being produced, it would intersect this curve and would not lie outside the curve on both sides [2.11], in contradiction to the supposition. Therefore the proposition has been established.

Theorem V

Proposition 5

If one chooses arbitrarily either on one and the same hyperbola or on opposite hyperbolas, two points and if one draws through them one straight line or two mutually parallel lines, then the rectangles contained by the parts of the line or lines drawn, cut off on both sides by the hyperbola and the asymptotes, are mutually equal [2.13].

Suppose that on one and the same hyperbola $BPCD$ or on opposite hyperbolas [2.12] $BPCD$ with asymptotes AE and AF two points B and C have been arbitrarily chosen and that two straight lines BD and CP have been drawn, mutually parallel, meeting the asymptotes at the points E, F, G, and H.[a] I say that the rectangle EBF is equal to the rectangle GCH.

Indeed, if one draws through these points B and C line segments BI, BL, CK, and CM, each parallel to one of the asymptotes and bounded by the other asymptote, then we have[1]–because the rectangles IBL and KCM[2] are equal, CM is to BL–i.e., CH is to BF–as IB is to KC, i.e.,[3] as EB is to GC,

---

[1]Prop. 3, Cor. 2 and 3, p. [183] and [184]; [a]this meeting point lies in the case of the first figure, anyway, within the angle EAF because of Prop. 4 on p. [184]; [1]VI, 16; [2]Prop. 3, p. [180]; [3]I, 29 and VI, 4.

Asymptotorum alterutri parallela esset, aut, si vel huic vel illi Asymptoto extra angulum E A F occurreret, ex puncto intra angulum K A H ad verticem ei qui Hyperbolam continet ducta hanc vel illam Asymptoton secaret; ideoque [1] curvæ in uno tantùm puncto non verò in duobus occurreret, ac producta eandem secaret, non autem utrinque extra Hyperbolam caderet, contra id quod ponitur. Ac proinde constat propositum.

[1] *per 2 & 3 Cor. 3 hujus.*

THEOREMA V.

Propositio 5.

Assumptis, vel in una eademque, vel in oppositis Hyperbolis, duobus utcunque punctis, ductisque per eadem sive unâ rectâ sive duabus, sibi mutuò parallelis: erunt rectangula sub ductæ vel ductarum partibus, Hyperbolâ & Asymptoto utrinque interceptis, sibi invicem æqualia.

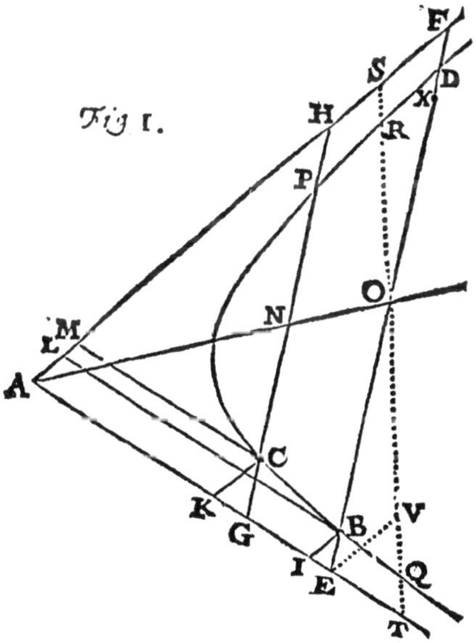

Fig. 1.

Sint, vel in eadem, vel in oppositis Hyperbolis B P C D, cujus Asymptoti A E, A F, assumpta utcunque bina puncta B & C, ac per eadem ductæ binæ rectæ B D, C P sibi invicem parallelæ Asymptotisque occurrentes in punctis E, F, G, H [4]: dico rectangulum E B F rectangulo G C H æquale esse.

[4] qui utique occursus in casu primæ figuræ fit intra angulum E A F, per 4 hujus.

Ductis enim per eadem puncta B & C rectis, utrique Asymptoto parallelis alteraque Asymptoto terminatis, B I, B L, C K, C M: erit [1], propter rectangula I B L & K C M [2] æqualia, ut I B ad K C, hoc est [3], ut E B

[1] *per 16 sexti.*

[2] *per 3 hujus.*

[3] *per 29 primi, & 4 sexti.*

[186] and therefore[4] the rectangles EBF and GCH are equal. This was to be proved. If through two points, let us say B and D, one straight line BD is drawn, meeting each of the asymptotes at the points E and F, then we can prove in the same way that the rectangles EBF and FDE are mutually equal.

Corollary 1

In the case of opposite hyperbolas the situation is as follows: if one of the parallel lines passes through the center, as CP in the third figure, we can show by means of the same proof that each of the rectangles, contained by the parts of arbitrary line segments drawn through the asymptotes to each of the two curves [2.14], equals the square on the parallel line segment drawn from the center to the hyperbola [2.15]. From the preceding it is clear that if an arbitrary straight line is drawn through the center, as CGP in the same figure, and if–wherever–another straight line, BD, is drawn parallel to it, intersecting the asymptotes at E and F, then the rectangle EBF or FDE will be equal to the square on GC as well as to the square on GP. From this it follows that GC and GP are also mutually equal, which means that an arbitrary line segment, drawn through the center to the opposite hyperbolas, is bisected at this center.

Corollary 2

It is also a fact that the parts of an arbitrary straight line, cut off by the hyperbola and by the asymptotes, are mutually equal, whether this straight line has been drawn through one and the same hyperbola or through opposite hyperbolas. Indeed, if BD has been drawn in an arbitrary way, intersecting the asymptotes at E and F, then it follows from the preceding[5] that BF is to DF as DE is to BE; therefore by separation,[6] or in opposite hyperbolas by composition[7] [2.16], BD will be to DF as the same BD is to BE; so DF and BE[8] and therefore BF and DE will be mutually equal.

Corollary 3

In the same way it follows that the straight line that joins two points -be it on one and the same hyperbola or on opposite hyperbolas- has no other point in common with the hyperbola.

---

[4]VI, 16; [5]Prop. 5, p. [185] and VI, 16; [6]V, 17; [7]V, 18; [8]V, 9.

4 per 16 sexti. ad GC, ita CM ad BL, id est, ita CH ad BF. ac proinde[4] rectangula EBF & GCH æqualia sunt. Quod demonstrandum erat.

Eodem modo ostendetur, si per bina puncta, ut B & D, una recta ducatur BD, quæ utrique Asymptoto occurrat in punctis E & F[a], rectangula EBF, FDE sibi invicem æqualia esse.

Corollarium 1.

In oppositis Hyperbolis, si parallelarum altera per centrum transeat, ut CP in tertia figura, eâdem demonstratione comprobatum erit, rectangula sub partibus quarumlibet rectarum, quæ per Asymptotos ad utramque curvam ducuntur, singula æqualia esse quadrato æquidistantis à centro ad Hyperbolam ductæ. Quare cum ex dictis appareat, si, ductâ per centrum rectâ utcunque veluti CGP in eadem figura, eidem ubivis alia recta æquidistans ducatur BD, quæ secet Asymptotos in E & F, rectangulum EBF vel FDE quadrato GC itemque & GP quadrato æquale esse: sequitur, ipsas quoque GC, GP esse sibi invicem æquales, hoc est, quamlibet rectam ad oppositas Hyperbolas per centrum ductam, in eodem centro bifariam secari.

Corollarium 2.

Constat quoque cujuslibet rectæ, sive per unam eandemque, sive per oppositas Hyperbolas ductæ, partes Hyperbolâ & Asymptotis interceptas sibi invicem esse æquales.

Ductâ enim utcunque BD, quæ Asymptotis occurrat in E & F, cum ex antedictis[5] BF sit ad DF, ut DE ad BE: erit quoque dividendo[6], vel, in oppositis Hyperbolis, componendo[7], BD ad DF, ut eadem BD ad BE, ideoque DF, BE[8], ac proinde & BF, DE sibi invicem æquales erunt.

5 per 5 hujus, & 16 sexti. 6 per 17 quinti. 7 per 18 quinti. 8 per 9 quinti.

Corollarium 3.

Unde pariter constat, rectam, quæ vel unius ejusdemque, vel oppositarum Hyperbolarum, bina puncta conjungit, nullo alio

sui.

[187] Indeed, if apart from D and B, any other point of DB, let us say X, should lie on the hyperbola, then[1] XF would be equal to BE, and so also equal to DF, and it would be as a part equal to the whole, which is absurd.

Corollary 4

It is, however, easily seen that the converse of this theorem is also true, viz., that, if with the same suppositions and the rectangles EBF and CGH being equal, one of the points B and C lies on the hyperbola, the other point will also lie on the same or on the opposite hyperbola, with asymptotes AE and AF. Indeed, because of the fact that the rectangles EBF and GCH are equal, it can be proved that the rectangles AIB and AKC are also equal, in the same way in which the converse has been proved before.

Therefore, if the point B lies on the hyperbola, the point C will also[2] lie on the same hyperbola or on the opposite hyperbola, whose asymptotes are AE and AF and vice versa. Regarding two points on the same line, e.g., B and D, the same can be said; the same will even be the case on the same line if the said points,

[1]by virtue of the preceding Cor.; [2]Prop. 3, p. [180].

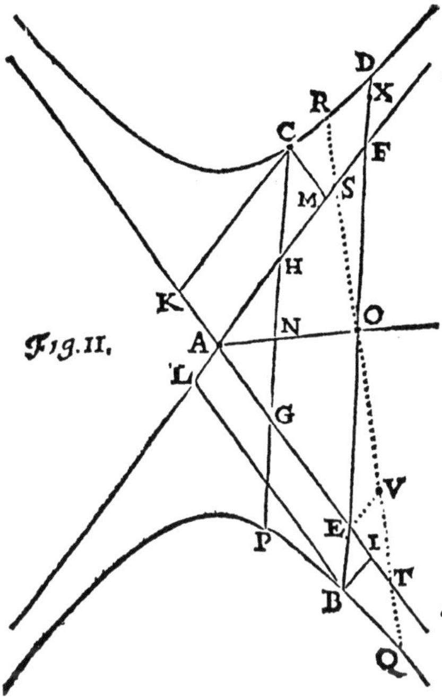

Fig. II.

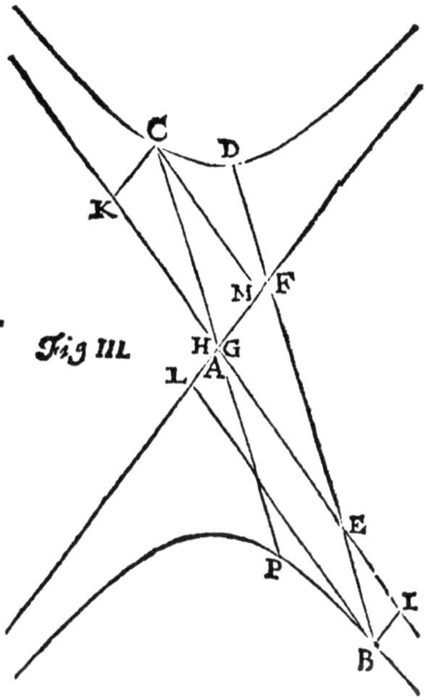

Fig III.

ſui puncto in Hyperbola eſſe. Si enim præter D & B aliud quoddam ipſius D B punctum, ex.gr. X, in Hyperbola foret, eſſet [1] X F ipſi B E ac proinde & ipſi D F æqualis, pars toti, quod eſt abſurdum. [1] per Coroll.præced.

Corollarium 4.

Facilè autem apparet, & converſum quoque propoſitionis verum eſſe: nempe, ſi, iiſdem poſitis, & rectangulis E B F, G C H æqualibus, punctorum B & C unum in Hyperbola ſit, & alterum quoque fore in eadem vel oppoſita Hyperbola, cujus Aſymptoti ſunt A E & A F. Ex eo enim quòd æqualia ſint rectangula E B F & G C H, demonſtrabitur æqualia quoque eſſe rectangula A I B & A K C eâdem methodo, quâ converſum ſupra oſtenſum fuit. ideoque ſi punctum B ſit in Hyperbola, erit quoque [2] punctum C in eadem aut in oppoſita Hyperbola, cujus Aſymptoti ſunt A E, A F, & vice verſâ. De binis autem punctis in eadem linea, ut B & D, idem dictum eſto; imò & idem erit in eadem linea, ſi dicta [2] per 3 hujus.

[188] let us say B and D, have the same distance to the asymptotes. Indeed, if BE and DF are equal and if we add to both BD or, in case of opposite hyperbolas, EF, then BF will be equal to DE and so the rectangle EBF will be equal to the rectangle FDE.

Corollary 5

It is also clear that a line through the center that bisects an arbitrary line segment, drawn in one and the same hyperbola or drawn between opposite hyperbolas, also bisects all line segments parallel to this line segment. E.g., let ANO, drawn from the center A, bisect the line segment CP, to which BD is parallel. Since NP and NC are equal, the line segments NH and NG will be equal too, after being added to them (i.e., NP and NC, transl.) or subtracted from them, the equal[1] segments PH and CG, and so OE and OF[2] are equal too. Likewise OB and OD are equal if again the equal line segments[3] BE and DF have been subtracted or added.

In this way, the straight lines drawn from the center through a hyperbola[a] are called the intercepted diameters or simply the diameters. The straight lines that are drawn from the center (in the region, transl.) between opposite hyperbolas,[b] are called the second diameters; the mutually parallel line segments that are bisected by these diameters[c] are said to be ordinate-wise applied to these diameters, and if the applied line segments are intersected by these diameters at right angles, then these diameters are called the axes of the hyperbola. When, however, the second diameter is parallel to the line segments that are ordinate-wise applied to the intercepted diameter, then we say that these diameters are mutually conjugate [2.17].

Corollary 6

From the preceding we conclude that there exist, apart from the said parallel line segments or ordinate-wise applied segments, no others that are bisected by a diameter. Suppose this is possible; let, distinct from the ordinate-wise applied line segments, another line segment, e.g., QR, be bisected by the diameter AO, intersecting the asymptotes at S and T, and through O let the line BOD be applied ordinate-wise, intersecting the asymptotes at E and F. Then it follows that EO and FO[4] are equal, as well as TO and SO.[5] If, however, we draw EV parallel to SF,[6]

[1]Prop. 5, Cor. 2, p. [186]; [2]V, 9 and VI, 4; [3] Prop. 5, Cor. 2, p. [186]; [a]like AO and the like in Fig. I; [b]like AO and the like in Fig. II. [c]like PC and DB in either figure; [4]Prop. 5, Cor. 2 and 5, p. [186] and p. [188]; [5]by hypothesis in Cor. 5, p. [188]; [6]I, 15 and 29.

puncta, ut B & D, æqualiter ab Aſymptotis diſtent: quandoquidem, ſi B E, D F æquales ſunt, additâ utrinque B D, vel in oppoſitis Hyperbolis ipsâ E F, & B F, ipſi D E, ideoque & rectangulum E B F rectangulo F D E æquale erit.

Corollarium. 5.

Apparet quoque, eam, quæ ex centro quamlibet rectam, vel in una eademque, vel in oppoſitis Hyperbolis ductam, bifariam dividit, omnes quoque ipſi æquidiſtantes bifariam dividere. Ut, ſi ex centro A ducta A N O dividat bifariam rectam C P, cui æquidiſtans ſit B D; cum, æqualibus N P, N C additis demptiſvè æqualibus [1] P H, C G, æquales quoque ſint N H, N G, ideoque, & O E, O F [2]: erunt ſimiliter, demptis rurſum additiſvè æqualibus [3] B E, D F, ipſæ O B & O D quoque æquales.

Ejuſmodi autem rectæ à centro per Hyperbolam ductæ [a], interceptæ diametri, ſeu diametri ſimpliciter; at quæ à cento inter oppoſitas Hyperbolas ducuntur [b], ſecundæ diametri; parallelæ verò per eaſdem bifariam ſectæ [c], ordinatim ad diametros applicatæ vocantur; & ſi applicatæ ad angulos rectos à diametris ſecentur, eædem diametri Hyperbolæ axes appellantur. Quando, autem ſecunda diameter ordinatim ad interceptam diametrum applicatis parallela eſt, altera alteri *Conjugata* dicitur.

[1] per 2 Cor. 5 hujus.
[2] per 9 quinti, & 4 ſexti.
[3] per 2 Cor. 5 hujus.
[a] ut A O ſimileſque in I figura.
[b] ut A O ſimileſque in II figura.
[c] ut P C, D B in utraque figura.

Corollarium 6.

Ex præmiſſis colligitur, non poſſe alias rectas, quàm dictas parallelas ſeu ordinatim applicatas, à diametro bifariam ſecari. Si enim fieri poſſit, ſecetur à diametro A O bifariam præter applicatas alia recta, ut Q R, Aſymptotis occurrens in S & T; & ſit per O ordinatim applicata B O D, Aſymptotis occurrens in E & F. Æquales ergo erunt tam E O, F O [4], quàm T O, S O [5]. Quoniam verò, ductâ E V ipſi S F parallelâ [6], æqui-

[4] per 2, & 5 Cor. 5 hujus.
[5] ex hypoth. juncto Cor. 5 hujus.
[6] per 15 & 29 Primi.

[189] then the triangles EOV and FOS are equiangular and[1] FO is to OS as EO is to OV. As EO is equal to FO, therefore OV also[2] equals OS, i.e., OT, and so it is as a part equal to the whole, which is absurd. Therefore the line segment QR is not bisected by the diameter AO.

Corollary 7

From this the following is clear: if in one and the same hyperbola or between two opposite hyperbolas two arbitrary mutually parallel line segments are drawn, then the straight line that bisects both of them will pass through the center, which means that it is a diameter.

Indeed, a diameter drawn through the midpoint of one of the parallel segments will also pass through the midpoint of the other.[3] From this it is clear how to find arbitrary diameters of a given hyperbola or of given opposite hyperbolas and at the same time the lines that are ordinate-wise applied to them, as well as the center, which is–as we know–the common point of intersection of two or more diameters.

---

[1]VI, 4; [2]V, 14; [3]Prop. 5, Cor. 5, p. [188].

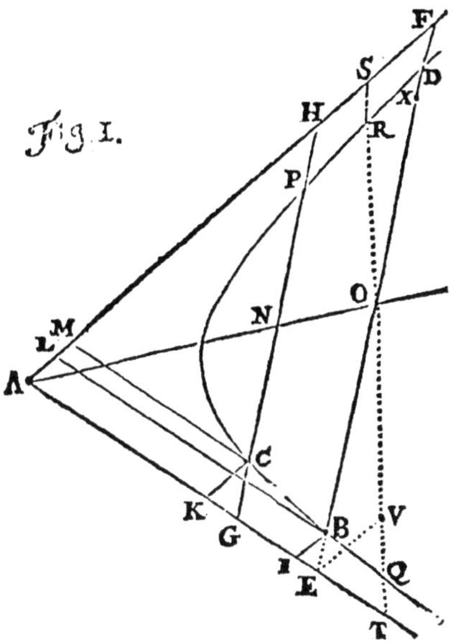

Fig. I.

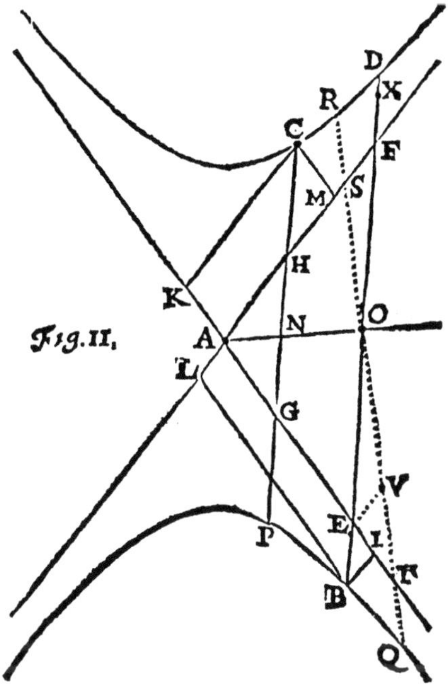

Fig. II.

æquiangula ſunt triangula E O V & F O S : erit [1] ut E O ad O V, ita F O ad O S. Quare cum E O ipſi F O ſit æqualis, erit &. [2] O V ipſi O S, hoc eſt, rectæ O T æqualis, pars toti, quod eſt abſurdum. Non ergo bifariam ſecatur recta R Q à diametro A O.

[1] per 4 ſexti.
[2] per 14 quinti.

Corollarium 7.

Atque hinc manifeſtum ſit, quòd, ſi vel in una eademque vel ad oppoſitas Hyperbolas binæ quælibet rectæ ſibi invicem æquidiſtantes ductæ ſint, quæ utramque bifariam dividit recta linea per centrum tranſeat ſeu diameter ſit: Quippe quæ per medium unius æquidiſtantium diameter ducetur, per medium quoque alterius æquidiſtantium tranſibit [3]. Unde apparet, quo pacto datæ Hyperbolæ vel oppoſitarum Hyperbolarum diametros quotlibet, ſimulque ordinatim applicatas ad eaſdem, nec non & centrum, utpote quod binarum pluriumvè diametrorum communis interſectio eſt, reperire liceat.

[3] per 5 Corol. 5 hujus.

[190]

Theorem VI

Proposition 6

A line segment that is drawn through an arbitrary point on a hyperbola to each of the asymptotes and that is bisected at this point, is a tangent to the curve at this point; conversely, a tangent to the curve will be bisected at this point, after being produced to the asymptotes.

Let the line segment GCH be drawn through the point C on the hyperbola BCD with asymptotes AE and AF, let it be bounded by the asymptotes on both sides, and let it be bisected at this point C. I say that the line segment GH is a tangent to the curve at C. Indeed, let–if possible–the line GH intersect the hyperbola at C and I: then[1] IH will be equal to the segment CG and therefore also equal to CH itself, which is absurd. So GH will not intersect the hyperbola, but will be a tangent to it. Furthermore I state the converse: if GH touches the hyperbola at the point C, then it is also bisected at C. Indeed, if this is not the case, then take on the greater part of CH a segment HI equal to GC. Since the point C lies on the hyperbola, the point I will also lie on the hyperbola[2] and the whole CI[3] will lie within the curve, and so GH will not touch the curve, but intersect it at the points C and I, contrary to the supposition. Therefore GC is not different from CH. So in both cases the proposition is true.

Corollary 1

In this way it is clear from the preceding that each of the rectangles contained by the parts of an arbitrary line segment, parallel to a tangent, and intercepted between the hyperbola and the asymptotes, equals the square on half the tangent. E.g., if BD has been arbitrarily drawn, parallel to the tangent GCH, meeting the asymptotes at E and F, then the rectangle EBF or[4] BFD, as well as FDE or DEB, will be equal to the rectangle GCH,[5] i.e., equal to the square on CH or on CG, each half the tangent.

---

[1]Prop. 5, Cor. 2, p. [186]; [2]Prop. 5, Cor. 4, p. [187]; [3]Prop. 5, Cor. 3, p. [186]; [4]Prop. 5, Cor. 2, p. [186]; [5]Prop. 5, p. [185].

THEOREMA VI.

Propositio 6.

Recta per quodlibet Hyperbolæ punctum ad utramque Asymptoton ducta, quæ in eodem puncto bifariam dividitur, curvam ibidem contingit; & contra, contingens ad utramque Asymptoton producta in puncto contactus bifariam divisa est.

Sit per punctum C in Hyperbola B C D, cujus Asymptoti A E, A F, ducta recta G C H, utrinque Asymptotis terminata, quæ in eodem puncto C bifariam dividatur. Dico rectam G H curvam contingere in C. Secet enim, si fieri potest, recta G H Hy-
[1] per 2 Cor. 5 hujus. perbolam in C & I: eritque [1] I H rectæ C G, ideoque & ipsi C H æqualis. quod est absurdum. Non secat ergo G H Hyperbolam, sed eandem contingit. Dico porrò conversim, si G H in puncto C Hyperbolam contingat, eandem quoque in C bifariam dividi. Hoc enim si non sit, sumatur in C H majori parte ipsa H I æqualis G C. Hinc cum punctum C sit in Hyperbola, erit quo-
[2] per 4 Cor. 5 hujus. [3] per 3 Cor. 5 hujus. que [2] punctum I in Hyperbola, totaque C I [3] intra curvam cadet, ideoque ipsa G H Hyperbolam non continget, sed eandem in punctis C & I secabit, contra id quod ponebatur. Non ergo G C ipsi C H inæqualis est. Ideoque casu utroque constat propositum.

Corollarium 1.

Manifestum itaque est ex antedictis, singula rectangula, quæ comprehenduntur sub partibus cujuslibet rectæ contingenti parallelæ, inter Hyperbolam & Asymptotos interceptis, esse æqualia dimidiæ tangentis quadrato. Ut, si tangenti G C H æquidistans utcunque ducta sit B D, Asymptotis occurrens in E & F:
[4] per 2 Cor. 5 hujus. [5] per 5 hujus. erit rectangulum E B F sive [4] B F D, ut & F D E sive D E B æquale rectangulo G C H [5], id est, ipsius C H vel C G, dimidiæ tangentis quadrato.

Co-

[191] Corollary 2

Moreover, it is clear that if a line segment is drawn through the endpoint of a diameter [2.20] and is bisected at the hyperbola by the same diameter, which means that it is parallel to the lines that are ordinate-wise applied to this diameter, then it touches the hyperbola at the said endpoint.

E.g., if BND has been ordinate-wise applied to the diameter AN, which (BND, transl.) being produced meets the asymptotes at E and F, and if through the endpoint C of the diameter the line segment GCH has been drawn parallel to BND, then–as NF and NE[1] are equal–CH and CG are also[2] equal and therefore[3] GCH will touch the hyperbola at C.

Corollary 3

From this it is clear not only that all line segments in a hyperbola that are parallel to a tangent are bisected by a diameter passing through the point of contact (and so are ordinate-wise applied to this diameter), but also that it is not possible that more lines can be tangent to a hyperbola at one and the same point. E.g., let BD be parallel to the tangent CH, meeting the asymptotes at E and F, and let the diameter ACN be drawn through the point of contact C, meeting BD in N; because[4] GC and CH are equal, as well as EN and NF,[5] BN and ND are also equal (after the equal[6] segments EB and DF are subtracted) and so they are ordinate-wise applied to the said diameter ACN. But it is clear that no straight line other than GH can be a tangent to the hyperbola at the point C, as all line segments drawn in the hyperbola parallel to this other line and which would be different from the previous ordinate-wise applied segments, would also be bisected by the same diameter,[7] which is not possible, as we established before.[8]

[1]Prop. 5, Cor. 2, p. [186] and Cor. 5, p. [188]; [2]V, 9 and VI, 4; [3]Prop. 6, p. [190]; [4]Prop. 6, p. [190]; [5]V, 9 and VI, 4; [6]Prop. 5, Cor. 2, p. [186]; [7]because of what has been proved above; [8]Prop. 5, Cor. 6, p. [188].

Corollarium 2.

Patet porrò, rectam, quæ per diametri terminum ducitur æquidistans ei, quæ in Hyperbola ab eadem diametro bifariam secatur, id est, ordinatim applicatis parallela, Hyperbolam in dicto termino contingere. Ut, si ad diametrum A N ordinatim applicata sit B N D, quæ producta Asymptotis occurrat in E & F, ac per diametri terminum C ducta sit recta G C H, ipsi B N D æquidistans, cum æquales sint N F & N E [1]: erunt [2] quoque C H & C G æquales, ideoque [3] G C H Hyperbolam continget in C.

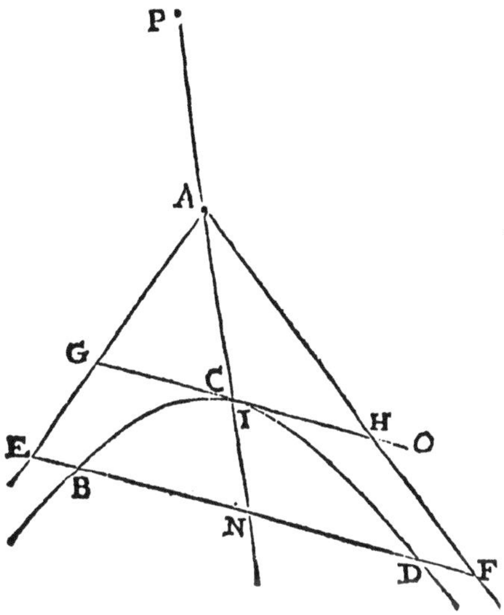

[1] *per 2 & 5 Corol. 5 hujus.*
[2] *per 9 quinti, & 4 sexti.*
[3] *per 6. hujus.*

Corollarium 3.

Hinc liquet, non solùm omnes rectas in Hyperbola, contingenti parallelas, à diametro per tactum ductâ bifariam secari, ideoque ad eam ordinatim applicatas esse, sed & non posse plures rectas in uno eodemque puncto Hyperbolam contingere. Ut, si contingenti C H parallela sit B D, Asymptotis occurrens in E & F, ductâ per tactum C diametro A C N, quæ ductæ B D occurrat in N: quoniam [4] G C, C H æquales sunt, nec non E N, N F [5], erunt quoque (demptis æqualibus [6] E B, D F,) B N, N D æquales, ideoque & ad dictam diametrum A C N ordinatim applicatæ. At verò non posse aliam rectam præter G H Hyperbolam in puncto C contingere, patet, quandoquidem & omnes ipsi æquidistantes in Hyperbola ductæ, quæque aliæ essent quàm prædictæ applicatæ, bifariam quoque per eandem diametrum dividerentur [7]. quod fieri non posse superiùs [8] ostensum est.

[4] *per 6 hujus.*
[5] *per 9 quinti & 4 sexti.*
[6] *per 2 Cor. 5 hujus.*
[7] *per supra demonstrata.*
[8] *in Cor. 6 5 hujus.*

Cæ-

[192] Moreover, it should be pointed out here–in order to define also the magnitude of the diameters–that the diameter drawn from an arbitrary point on a hyperbola through the center and bounded by the opposite hyperbola and therefore twice[1] the segment intercepted between the center and the curve, as CAP, is called the transverse diameter of the hyperbola or of the opposite hyperbolas. The line segment that touches the curve at the endpoint of the diameter mentioned before and that is bounded by the asymptotes on both sides, or the line segment–equal and parallel to it, as GCH–that passes through the center, is called the second diameter, conjugate to the transverse diameter. The line segment that is the third proportional to PC and GH, viz., to the transverse and the second diameter, as CO, is called the latus rectum or the parameter [2.21].

Theorem VII

Proposition 7

A straight line drawn through the endpoint of an arbitrary transverse diameter, parallel to the tangent at the top, touches the opposite hyperbola, and a straight line that is ordinate-wise applied to a second diameter that is conjugate to an arbitrary diameter, is parallel to this assumed diameter.

[1] Prop. 5, Cor. 1, p. [186].

Cæterùm monendum hîc, ut diametrorum quoque magnitudo determinetur, eam, quæ à quocunque in Hyperbola puncto per centrum ducta oppositâ Hyperbolâ terminatur, ideoque interceptæ inter centrum & curvam dupla est [1], ut CAP, vel Hyperbolæ, vel oppositarum Hyperbolarum transversam; diametrum; eamque, quæ in ipsius termino curvam contingens utrinque Asymptotis terminatur, aut quæ ipsi per centrum æqualis & parallela ducitur, ut GCH, secundam diametrum transversæ conjugatam; at verò illam, quæ ipsis PC, GH, transversæ nempe secundæque diametro tertia est proportionalis, ut CO, rectum latus sive Parametrum dici.

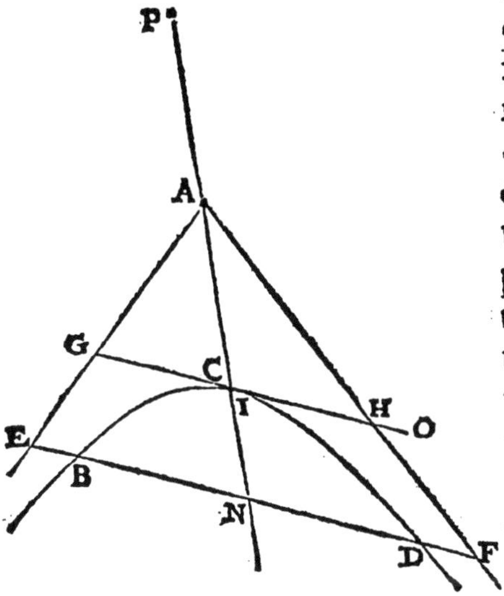

[1] per Corol. 1 5 hujus.

THEOREMA VII.

Propositio 7.

Quæ per terminum transversæ cujuslibet diametri recta ducitur, contingenti in vertice parallela, oppositam Hyperbolam contingit, & quæ ad secundam diametrum, assumptæ cuicunque diametro conjugatam, ordinatim applicatur, eidem assumptæ diametro æquidistat.

Sit

[193] In the hyperbola IC or in the opposite hyperbolas IC and HC, with asymptotes BG and DF, let there be chosen an arbitrary transverse diameter CE, and through its endpoint E let there be drawn the straight line FEG, parallel to BD, which (BD, transl.) touches the curve at C in such a way that the latter and the former meet the asymptotes at the points B, D and F, G, respectively.

I say that the said FEG also touches the hyperbola, viz., at E. Moreover, if one draws a second diameter AK through the center A, conjugate to the diameter CE, then the straight lines that are ordinate-wise applied to AK are parallel to the diameter CE.

Indeed, because[1] AE is to EG as AC is to CB and AE is to EF as AC is to CD and because AE and AC[2] are equal, as well as CB and CD,[3] therefore EG also[4] equals CB and EF equals CD and so EG equals EF. For this reason[5] the straight line FG will touch the hyperbola at the point E. This was the first proposition.

Furthermore the following is true [2.22]: if one draws the straight line GD through G and D, which intersects the second diameter AK at K and meets the opposite hyperbolas at H and I, then it is known that EG and CD are equal and parallel and so[6] the lines GD and CE that connect them, are also parallel and equal. As the second diameter AK is parallel to the tangents BD and FG (i.e.,[7] parallel to lines that are ordinate-wise applied to the diameter CE)–for AK is conjugate to CE, as supposed–therefore the lines GK and EA as well as KD and AC are also[8] equal and so GK and KD are also equal. If we add to these the equal[10] segments GH and DI, then the segments KH and KI will also be mutually equal. As[11] the segment HI has been ordinate-wise applied to the second diameter AK, therefore all other straight lines that are ordinate-wise applied to the same diameter,[12] will also be parallel to HI and so they are also parallel to the diameter CE. This was the second proposition.

---

[1] I, 29 and VI, 4; [2] Prop. 5, Cor. 1, p. [186]; [3] Prop. 6, p. [190]; [4] V, 14; [5] Prop. 6, p. [190]; [6] I, 33; [7] Prop. 6, Cor. 3, p. [191]; [8] I, 34; [9] Prop. 5, Cor. 1, p. [186]; [10] Prop. 5, Cor. 2, p. [186]; [11] Prop. 5, Cor. 6, p. [188]; [12] Prop. 5, Cor. 5 and 6, p. [188].

Sit Hyperbolæ, vel oppoſitarum Hyperbolarum I C, H E, quarum Aſymptoti B G, D F, diameter tranſverſa utcunque aſſumpta C E, perque ejus terminum E ducta recta F E G parallela ipſi B D, quæ curvam in vertice C contingit, ita ut hæc atque illa Aſymptotis occurrant in punctis B, D & F, G: dico prædictam quoque F E G oppoſitam Hyperbolam contingere in E; & ſi per centrum A ducatur ſecunda diameter A K, diametro C E conjugata, ordinatim ad eandem A K applicatas ipſi C E diametro æquidiſtare.

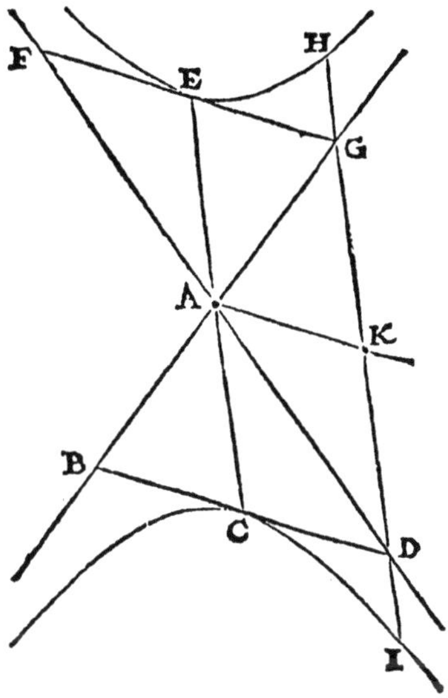

Quoniam enim eſt [1] tam A E ad E G, ut A C ad C B, quàm A E ad E F, ut A C ad C D; & ſunt tam A E, A C [2] quàm C B, C D [3] æquales, erit quoque [4] tam E G ipſi C B, quàm E F ipſi C D, ac proinde & E G ipſi E F æqualis. Unde [5] recta F G oppoſitam Hyperbolam H E continget in puncto E. Quod primo loco propoſitum fuit. Porrò ſi per G & D ducatur recta G D, ſecans ſecundam diametrum A K in K, oppoſitisque Hyperbolis occurrens in H & I, cum æquales & parallelæ ſint E G, C D, erunt & [6] quæ ipſas conjungunt G D, C E parallelæ & æquales. Ideoque cum ſecunda diameter A K contingentibus B D, F G, id eſt [7] ordinatim ad diametrum C E applicatis æquidiſtans ſit, utpote ex Hypotheſi ipſi C E conjugata: erunt quoque [8] rectæ G K, E A, ut & K D, A C, ideoque & [9] G K, K D æquales. Quibus ſi addantur æquales [10] G H, D I: erunt ſimiliter rectæ K H, K I ſibi invicem æquales. Quocirca cum [11] ad ſecundam diametrum A K applicata ſit recta H I, etiam cæteræ omnes ad eandem applicatæ [12] eidem H I ac proinde & diametro C E æquidiſtabunt. Quod ſecundo loco propoſitum erat.

[1] per 29 primi, & 4 ſexti.
[2] per 1 Corol. 5 hujus.
[3] per 6 hujus.
[4] per 14 quinti.
[5] per 6 hujus.
[6] per 33 primi.
[7] per 3 Cor. 6 hujus.
[8] per 34 primi.
[9] per 1 Cor. 5 hujus.
[10] per 2 Cor. 5 hujus.
[11] per 6 Cor. 5 hujus.
[12] per 5 & 6 Cor. 5 hujus.

[194]

Problem I

Proposition 8

To find the conjugate axes of a hyperbola if arbitrary conjugate diameters are given.

Let PC and GH be the given conjugate diameters [2.23] and let it be required to find the conjugate axes of the hyperbola of which PC and GH are conjugate diameters.

First we draw from the center A through G and H the asymptotes AG and AH, and from C we draw the line segment CB to one of the asymptotes, parallel to the other, and we take the mean proportional AD between AB and BC. Next we draw DE equal to AD and parallel to the asymptote AH; then EAF, passing through E and A and twice EA, will be the sought transverse axis and IEK, perpendicular to it and bounded by each of the asymptotes on both sides, will be the second diameter conjugate to the former [2.24].

Indeed, as the point C[1] lies on the hyperbola and as the rectangle ADE equals the rectangle ABC,[2] therefore the point E[3] will also lie on the hyperbola. Furthermore: as the line segments DA and DE are equal,[4] so the angle DAE equals the angle DEA, i.e.,[5] the angle EAK, and as the angles AEI and AEK are also equal, by construction, therefore[6] the triangles AEI and AEK are equiangular; because they have the side AE in common,[7] they are equal and so the side IE equals the side EK. As the point E[8] lies on the hyperbola and bisects the segment IK (bounded by the hyperbola on both sides), therefore IK[9] itself will be tangent to the curve at E: for this reason, and because the angles FEI and FEK are right angles, FE and IK will be the conjugate axes.

---

[1]by hypothesis; [2]VI, 17; [3]Prop. 3, p. [180]; [4]I, 5; [5]I, 29; [6]I, 32; [7]I, 26; [8]by virtue of what has been proved above; [9]Prop. 6, p. [190].

PROBLEMA I.

Propoſitio 8.

Datis quibuſcunque diametris conjugatis, Hyperbolæ axes conjugatos invenire.

Sint datæ diametri conjugatæ P C, G H, oporteatque invenire conjugatos axes ejus Hyperbolæ, cujus eædem P C, G H conjugatæ diametri exiſtunt.

Ductis ab A centro per G & H Aſymptotis A G, A H, ductâque à C ad eorum alterutram rectâ C B alteri æquidiſtante, ſumatur inter A B, B C media proportionalis A D. Dein ductâ D E ipſi A D æquali, atque Aſymptoto A H parallelâ, erit E A F, tranſiens per E & A ac ipſius E A dupla, tranſverſus axis qui quæritur, atque I E K ad eandem perpendicularis, ac utrinque Aſymptotis terminata, axis ſecundus, priori conjugatus.

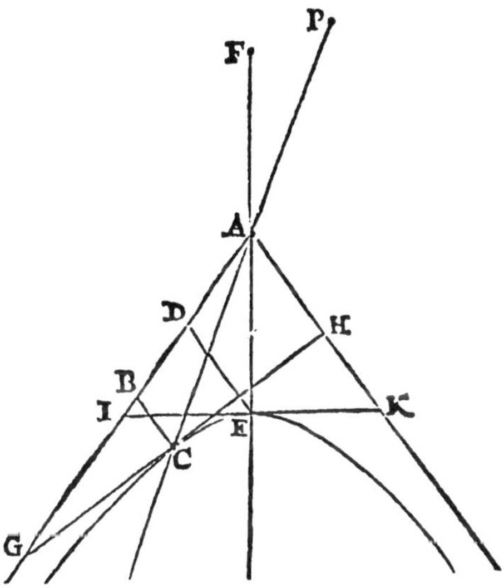

Quoniam enim punctum C [1] in Hyperbola eſt, rectangulumque A D E ipſi A B C æquale [2]; erit quoque punctum E [3] in Hyperbola. Porrò cum propter rectas D A, D E æquales [4] æqualis quoque ſit D A E angulus ipſi D E A, id eſt [5], E A K angulo, ſintque & anguli A E I, A E K ex conſtructione æquales: erunt [6] triangula A E I, A E K æquiangula, atque ob latus A E commune [7] etiam æqualia, latusque I E lateri E K æquale. Unde cum punctum E [8] in Hyperbola exiſtat, dividatque bifariam rectam I K, utrinque Aſymptotis terminatam, continget ipſa I K [9] curvam in E: ideoque, & propter angulos F E I, F E K rectos, conjugati axes erunt F E, I K.

[1] *ex hypotheſi.*
[2] *per 17 ſexti.*
[3] *per 3 hujus.*
[4] *per 5 primi.*
[5] *per 29 primi.*
[6] *per 32 primi.*
[7] *per 26 primi.*
[8] *per ſup. demonſtr.*
[9] *per 6 hujus.*

THEO-

[195]

Theorem VIII

Proposition 9

Arbitrary tangents cut off equal triangles from the angle contained by the asymptotes of a hyperbola, and the rectangles comprehended by the sides of these triangles are also mutually equal. Besides, their slant sides are divided by the tangents in the same ratio; their bases, i.e., the tangents bounded by the asymptotes, are divided in the same ratio by their point of intersection; the parts of these bases that touch the curve and that lie between the point of intersection and the asymptotes, are likewise divided in the same proportion by the point of contact [2.25].

Let the straight lines *GH* and *IK* (bounded by the asymptotes of the hyperbola *CE* with asymptotes *AG* and *AK* on both sides and meeting each other at *R*) be tangent to the hyperbola at the points *C* and *E*. I say that both the rectangles and the triangles *GAH* and *IAK* are equal. Moreover, *GI* is to *IA* as *KH* is to *HA*; likewise, *GR* is to *RH* as *KR* is to *RI* and *GC* is to *CR* as *KE* is to *ER*. Indeed, first draw from the points of contact, *C* and *E*, the straight lines *CB* and *ED* parallel to one of the asymptotes, e.g., *AH*. *GB* is to *GA* as *GC* is to *CH* and as *BC* is to *AH*,[1] and *GH* is twice *GC*;[2] therefore *GA* will also be twice *GB* and *AH* will be twice *BC*, so[3] the rectangle *GAH* will be four times the rectangle *GBC* or *ABC*. In the same way we can show that the rectangle *IAK* is four times the rectangle *ADE*. As the rectangles *ABC* and *ADE* are equal,[4] so their quadruples, namely *GAH* and *IAK*, are equal. This is the first statement. Because[5] *IA* is to *AH* as *GA* is to *AK*, therefore the triangles *GAH* and

[1]VI, 4; [2]Prop. 6, p. [190]; [3]VI, 20; [4] Prop. 3, p. [180]; [5]VI, 16.

THEOREMA VIII.

Propoſitio 9.

Quælibet contingentes ab angulo Hyperbolæ Aſymptotis comprehenſo æqualia abſcindunt triangula, & rectangula ſub eorundem triangulorum lateribus comprehenſa invicem quoque æqualia ſunt, ac præterea majora eorundem latera à contingentibus, ipſæque baſes ſeu contingentes Aſymptotis terminatæ, in mutuo occurſu, nec non ipſarum partes curvam contingentes inter occurſum & Aſymptotos interjectæ, in punctis contactus, in eadem ratione ſecantur.

Hyperbolam C E, cujus Aſymptoti A G, A K, rectæ G H, I K utrinque Aſymptotis terminatæ, ac ſibi mutuò in R occurrentes, contingant in punctis C & E: dico tam rectangula quàm triangula G A H, I A K æqualia eſſe; ac præterea eſſe G I ad I A, ſicut K H ad H A; itemque G R ad R H, ſicut K R ad R I; nec non G C ad C R, ſicut K E ad E R.

Ductis enim à punctis contactus C & E rectis C B, E D Aſymptotorum alterutri, ut A H, parallelis, cum ſit ut G C ad G H, ita G B ad G A, & B C ad A H [1]; ſitque G H ipſius G C dupla [2]. [1] per 4 ſexti. [2] per 6 hujus.

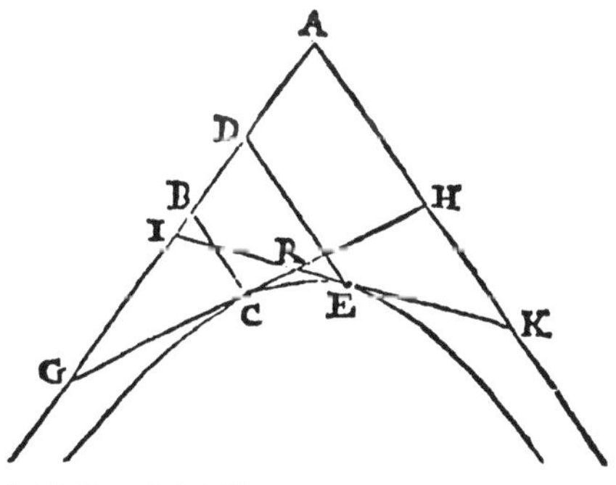

erit quoque tam G A ipſius G B quàm A H ipſius B C dupla, ideoque [3] rectangulum G A H rectanguli G B C ſive A B C quadruplum. Eodem modo rectangulum I A K rectanguli A D E quadruplum oſtendetur. Hinc cum æqualia ſint rectangula A B C, A D E [4], erunt quoque eorum quadrupla, nimirum rectangula G A H & I A K æqualia. Quod eſt primum. [3] per 20 ſexti. [4] per 3 hujus.

Unde cum [5] ſit ut G A ad A K, ita I A ad A H, triangula quo- [5] per 16 ſexti.

[196] IAK will also[6] be equal, as the sides at the angle that they have in common are inversely proportional. This is the second statement.

By interchanging the means,[7] GA is also to IA as AK is to AH; therefore, by separation[8] [2.16], GI will be to IA as KH is to HA. This is the third statement.

If we subtract from the equal triangles GAH and IAK the common quadrilateral $IRHA$, the remainders, viz., the triangles GRI and $KRHI$, will also be equal and therefore[9] their sides at the equal angle at R will be inversely proportional, which means that GR is to RH as KR is to RI. This is the fourth statement.

Also, by composition[10] [2.16] GH is to RH as KI is to RI or –if we take halves–CH is to HR as EI is to IR; therefore CH or GC will be to CR as EI or KE is to ER (by conversion of the ratio[11] [2.25]). This is the fifth statement, and so all statements have been proved.

Theorem IX

Proposition 10

If an arbitrary diameter has been drawn in a hyperbola (see the figure on p. [198], transl.), then the square on an arbitrary ordinate-wise applied segment will be to the rectangle contained by the parts of this diameter that are cut off between its endpoints and the the applied line, as the square of the second diameter is to the square of the transverse diameter, i.e., as the parameter is to the transverse diameter [2.26].

In the hyperbola BCD with asymptotes AE and AF let there be drawn an arbitrary diameter $PACN$, with second diameter GCH–conjugate to the transverse diameter PC–and with parameter CI, which is–as we know–the third proportional to PC and GH, and let an arbitrary line segment ND be ordinate-wise applied to the said diameter DN.

[6]VI, 15; [7]V, 16; [8]V, 17; [9]VI, 15; [10]V, 18; [11]V, 19, Cor.

196 ELEM. CURVARUM

6 per 15 ſexti. que G A H, I A K æqualia erunt[6], utpote habentia latera circa communem angulum, reciproca. Quod eſt ſecundum.

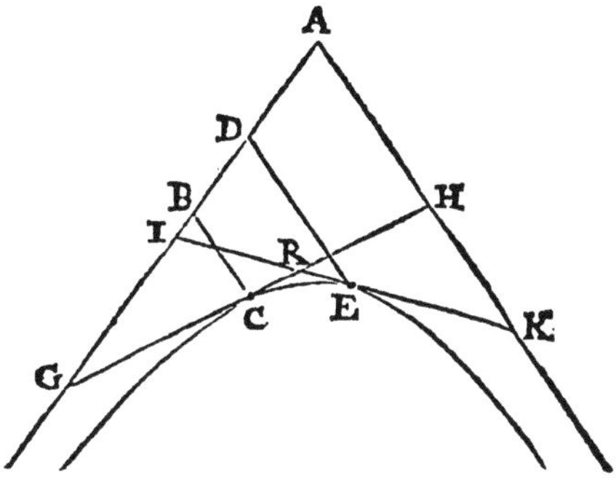

7 per 16 quinti. 8 per 17 quinti. Ac cum permutando [7] quoque ſit G A ad I A, ut A K ad A H: erit & [8] dividendo G I ad I A, ut K H ad H A. Quod eſt tertium.

9 per 15 ſexti. Porrò cum ab æqualibus triangulis G A H, I A K ablato communi quadrilatero I R H A, reſidua, nempe triangula G R I & K R H, quoque æqualia remaneant, erunt [9] eorundem latera circa æqualem angulum ad R reciproca, id eſt, erit G R ad R H, ut K R ad R I. Quod eſt quartum.

10 per 18 quinti. 11 per Coroll. 19 quinti. Unde cum componendo [10] quoque ſit G H ad R H, ut K I ad R I, aut, ſumptis antecedentium dimidiis, C H ad H R, ut E I ad I R: erit & per converſionem rationis [11] C H ſive G C ad C R, ut E I ſive K E ad E R. Quod eſt quintum. Atque ita demonſtrata ſunt ea, quæ proponebantur.

THEOREMA IX.

Propoſitio 10.

Ductâ quacunque in Hyperbola diametro, erit ut quadratum ſecundæ ad quadratum tranverſæ diametri, ſive ut parameter ad tranſverſam diametrum, ita quadratum cujuſlibet ordinatim applicatæ ad rectangulum ſub ejuſdem diametri partibus, utroque tranſverſæ termino & applicatâ interceptis, comprehenſum.

Sit in Hyperbola B C D, cujus Aſymptoti A E, A F, ducta diameter utcunque P A C N, cujus ſecunda diameter tranſverſæ P C conjugata ſit G C H, parameter verò C I, ipſis nempe P C, G H tertia proportionalis, & ſit ordinatim ad dictam diametrum appli-

[197] I say that the square on DN is to the rectangle PCN as the square on GH is to the square on CP, or–which is the same[1]–as the segment IC to the segment CP.

Indeed, if we produce the applied DN, on both sides, through the hyperbola to the asymptotes, e.g., as $EBNDF$, then the following is true: because the square on FN[2] is to the square on HC–i.e.,[3] to the rectangle BFD– as the square on NA is to the square on CA, therefore (by separation[4] [2.16]), the square on DN will be to the square on HC as the rectangle PNC[5] is to the square on CA and, by interchanging the means,[6] the square on DN is to the rectangle PNC as the square on HC is to the square on CA or[7] as the square on GH is to the square on CP, or–which is the same–as IC is to CP. This was to be proved.

Corollary 1

From this it follows how to find the asymptotes of an arbitrarily given hyperbola, e.g., BCD. In fact, first we construct[8] [2.27], the center A, an arbitrary diameter AN intersecting the curve at C, and a line segment that has been ordinate-wise applied to it. Then the following is true: if we produce NA to P in such a way that AP equals AC, and if we draw a line CGI through C parallel to the applied BN and if we mark on it the points H and G so that the rectangle PNC is to the square on BN as the square on AC is to the square on CG or CH, then the straight lines AGE and AHF, drawn from the center A through G and H, will be the required asymptotes[9].

Corollary 2

From what we proved, the following is clear: if we draw the line PIK through P and I, the endpoints of a transverse diameter and of the (corresponding, transl.) parameter respectively,

---

[1] VI, 20, Cor.; [2] VI, 4 and 22; [3] Prop. 6, Cor. 1, p. [190]; [4] II, 6 and V, 17; [5] II, 6; [6] V, 16; [7] V, 15; [8] Prop. 5, Cor. 7, p. [189]; [9] by the converse of Prop. 10, p. [196].

applicata quælibet D N: dico eſſe ut G H quadratum ad C P quadratum, aut, quod idem eſt [1], ut recta I C ad rectam C P, ita quadratum D N ac P N C rectangulum.

Productâ enim applicatâ D N utrinque per Hyperbolam ad Aſymptotos, ut E B N D F, cum ſit [2] F N quadratum ad H C quadratum, id eſt [3], ad B F D rectangulum, ut N A quadratum ad C A quadratum: erit dividendo [4] D N quadratum ad H C quadratum, ut P N C rectangulum [5] ad C A quadratum, & permutando [6] D N quadratum ad P N C rectangulum, ut H C quadratum ad C A quadratum, ſive [7] ut G H quadratum ad C P quadratum, aut, quod idem eſt, ut I C ad C P. Quod demonſtrandum erat.

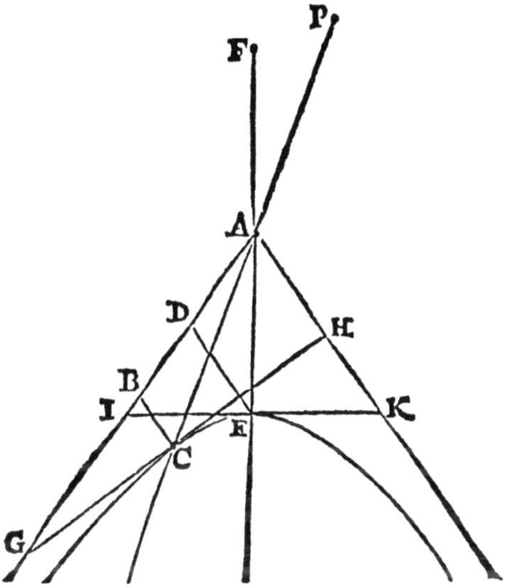

Corollarium 1.

Hinc colligitur, quo pacto datæ cujuſlibet Hyperbolæ, ut B C D, Aſymptoti inveniantur. Quippe inventis [8] centro A, diametro quâcunque A N quæ curvam ſecet in C, & ordinatim ad eandem applicatâ B N; ſi, productâ N A ad P, ut A P ipſi A C ſit æqualis, ductâque per C rectâ G C I applicatæ B N parallelâ, in eadem notentur puncta H & G, ita ut ſit P N C rectangulum ad B N quadratum, ſicut A C quadratum ad quadratum abs C G ſeu C H: erunt, quæ ex A centro per G & H ducuntur rectæ A G E & A H F, Aſymptoti quæſitæ [9].

Corollarium. 2.

Ex demonſtratis patet, ſi per P & I tranſverſæ diametri parametrique terminos ducatur recta P I K, occurrens cuilibet appli-

1 *per Corol. 20 ſexti.*
2 *per 4, & 22 ſexti.*
3 *per 1 Corol. 6 hujus.*
4 *per 6 ſecundi, & 17 quinti.*
5 *per 6 ſecundi.*
6 *per 16 quinti.*
7 *per 15 quinti.*
8 *per 7 Coroll. 5 hujus.*
9 *per converſum 10 hujus.*

[198] which (PIK, transl.) meets an arbitrary applied straight line, e.g., ND, being produced if necessary, at K, then the rectangle CNK equals the square on the applied line segment DN [2.28].

Indeed,[1] the rectangle PNC is to the square on DN as PC is to CI or as PN is to NK,[2] i.e., (if we take NC as the common altitude), as the rectangle PNC is to the rectangle CNK.[3] Therefore[4] the rectangle CNK[4]will be equal to the square on the applied DN.

This means, if we may say it in the way of the ancient geometers [2.29]:

> *The line segment that has been ordinate-wise applied to a diameter from a point on a hyperbola, has–if squared–the same area as the rectangle adjacent to the latus rectum, with as breadth the segment that is cut off from the diameter between the applied segment itself and the vertex of the diameter, but exceeding this (the area of this rectangle, transl.) by a figure that is similar to and that has the same position as the figure contained by the transverse diameter and the latus rectum.*

Corollary 3

From what we proved, it is also clear that in a hyperbola the squares on ordinate-wise applied line segments have the same ratio as the rectangles contained by those parts that are cut off from the diameter, measured from each of the endpoints of the transverse diameter. E.g., if the applied line segments are LM and DN, then the square on DN will be to the rectangle PNC as the square on LM is to the rectangle PMC, because in both cases the ratio is the same as the ratio of the parameter to the transverse diameter.[5] Therefore[6]–by interchanging the means–the square on LM will be to the square on DN as the rectangle PMC is to the rectangle PNC.

---

[1]by the converse of Prop. 10, p. [196]; [2]VI, 4; [3]VI, 1; [4]V, 9; [5]Prop. 10, p. [196]; [6] V, 16.

catæ, ut N D, productæ, si opùs fuerit, in K: rectangulum C N K quadrato applicatæ D N æquale esse. Quoniam enim est [1] ut P C ad C I, sive ut P N ad N K [2], id est, (sumptâ N C communi altitudine) ut P N C rectangulum ad C N K rectangulum [3], ita idem P N C rectangulum ad D N quadratum, erit [4] rectangulum C N K quadrato applicatæ D N æquale, id est, si veterum Geometrarum more id proponi placeat:

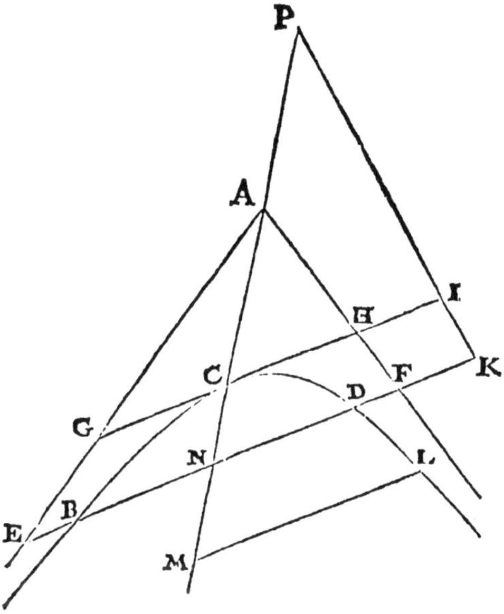

[1] *per 10 hujus conv.*
[2] *per 4 sexti.*
[3] *per 1 sexti.*
[4] *per 9 quinti.*

Quæ ab Hyperbola ad diametrum ordinatim applicatur, potest spatium adjacens lateri recto, latitudinem habens lineam, quæ à diametro abscinditur inter ipsam applicatam & diametri verticem interjectam, excedensque figurâ simili similiterque positâ ei, quæ lateribus transverso rectoque continetur.

Corollarium 3.

Manifestum quoque est ex demonstratis, in Hyperbola applicatarum quadrata ad se invicem esse, veluti rectangula sub interceptis diametri portionibus, ab utroque transversæ termino sumptis, ut, si applicatæ sint L M, D N, erit ut quadratum L M ad rectangulum P M C, ita quadratum D N ad rectangulum P N C; cum utriusque eadem sit ratio, quæ est parametri ad transversam diametrum [5], eritque propterea [6] permutatim L M quadratum ad D N quadratum, ut P M C rectangulum ad P N C rectangulum.

THEO-

[5] *per 10 hujus.*
[6] *per 16 quinti.*

[199]

Theorem X

Proposition 11

If an arbitrary tangent intersects an arbitrary diameter of a hyperbola and if from the point of contact a line segment is ordinate-wise applied to this diameter, then the rectangle contained by the parts of the diameter that are cut off by the tangent and the applied segment–measured from the center–will be equal to the square on half the diameter [2.30].

Suppose that the straight line ECF touches an arbitrary hyperbola KC, with asymptotes AD and AF, at an arbitrarily chosen point C, and suppose that it intersects the asymptotes at E and F, but an arbitrary diameter AH at I. Suppose moreover that through the point of contact C the line segment CH has been ordinate-wise applied to this diameter, intersecting–being produced–an asymptote at M. I say that the rectangle HAI will be equal to the square on half the diameter KA, or–which is the same[1]–that HA, KA and IA are continuedly proportional.

Indeed, if we draw DKG parallel to the applied segment CH and KL parallel to the tangent FE, and if we denote the point of intersection (of DKG and ECF, transl.) by R, then the following is true: because RC is to CF[2] as RK is to KD, which means[3] that MG is to MF as LE is to LD, therefore[4] MG is also to GF as LE is to ED. Moreover,[5] as FG is to GA as DE is to EA, therefore (ex aequo[6] [2.30]) MG will be to GA, i.e.,[7] HK will be to KA, as LE is to EA, which means[8]: as KI is to IA and by composition[9] [2.16] HA will be to KA as KA is to IA. Which was to be proved.

---

[1]VI, 17; [2]VI, Cor. 2 and Prop. 9, p. [195]; [3] VI, 2; [4]by composition of the ratios, see Clavius's commentary on V, 18; [5]Prop. 9, p. [195]; [6]V, 22; [7]VI, 2; [8] VI, 2; [9]V, 18.

THEOREMA X.

Propositio 11.

Si quælibet contingens cuicunque Hyperbolæ diametro occurrat, atque à puncto contactus recta ad eandem diametrum ordinatim applicetur, erit rectangulum sub diametri portionibus à centro per contingentem applicatamque abscissis æquale semidiametri transversæ quadrato.

Quamcunque Hyperbolam K C, cujus Asymptoti A D, A F, contingat in puncto C utcunque sumpto recta E C F, Asymptotis occurrens in E & F, diametro autem A H utcunque ductæ in I; & per punctum contactus C ad eandem diametrum ordinatim applicata sit C H, quæ producta Asymptoto occurrat in M. Dico rectangulum H A I æquale fore quadrato semidiametri K A, sive, quod idem est [1], continuè proportionales esse H A, K A, & I A.

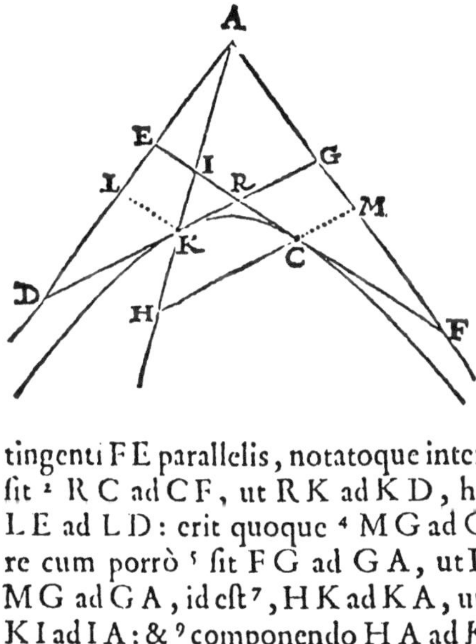

Ductis enim D K G applicatæ C H, & K L contingenti F E parallelis, notatoque intersectionis puncto R, cum sit [2] R C ad C F, ut R K ad K D, hoc est [3], M G ad M F, ut L E ad L D: erit quoque [4] M G ad G F, ut L E ad E D. Quare cum porrò [5] sit F G ad G A, ut D E ad E A: erit [6] ex æquo M G ad G A, id est [7], H K ad K A, ut L E ad E A, hoc est [8], ut K I ad I A: & [9] componendo H A ad K A, ut K A ad I A. Quod demonstrandum erat.

[1] *per 17 sexti.*
[2] *per 2 Cor. sexti, & 9 hujus.*
[3] *per 2 sexti.*
[4] *per compositionem rationis contrariam, vide Clavium ad 18 quinti.*
[5] *per 9 hujus.*
[6] *per 22 quinti.*
[7] *per 2 sexti.*
[8] *per 2 sexti.*
[9] *per 18 quinti.*

THEO-

[200]

Theorem XI

Proposition 12

If an arbitrary tangent intersects an arbitrary second diameter of a hyperbola and if from the point of contact a line segment is ordinate-wise applied to this diameter, then the rectangle contained by the parts of the second diameter that are cut off by the tangent and the applied segment- measured from the center- will be equal to the square on half the second diameter [2.31].

Suppose that the straight line FCQ touches an arbitrary hyperbola KC, with asymptotes AD and AF, at an arbitrarily chosen point C, and suppose that it intersects an arbitrarily chosen second diameter AB at Q. Then I say: if from C the line segment CB is ordinate-wise applied to this diameter AB and if from A, parallel to CB, the straight line AKH is drawn, intersecting the tangent FCQ at I and meeting the hyperbola at K, and if through K the straight line DKG is drawn, parallel to AB (so that[1] AKH is conjugate to the second diameter AB and so that half the second diameters have the length KG and KD), then the rectangle BAQ will be equal to the square on half the diameter KG or KD.

Indeed, if we draw through C the straight line TCM, parallel to the second diameter AB and thus ordinate-wise applied to the transverse diameter AKH, which (TCM, transl.) intersects the hyperbola at T and meets the diameter AH at H, the asymptote AF at M, then the following is true: because[2] the square on HA is to the square on KA, or[3] the square on HM is to the square on KG

[1]Prop. 7, p. [192]; [2]Prop. 11, p. [199] and VI, 20, Cor.; [3]VI, 4 and 22.

THEOREMA XI.

Propoſitio 12.

Si quælibet contingens cuicunque ſecundæ Hyperboles diametro occurrat, atque à puncto contactus recta ad eandem diametrum ordinatim applicetur, erit rectangulum ſub ſecundæ diametri portionibus, à centro per contingentem applicatamque abſciſſis, æquale ſemi-ſecundæ diametri quadrato.

Quamcunque Hyperbolam K C, cujus Aſymptoti A D, A F, contingat in puncto C, utcunque ſumpto, recta F C Q, occurrens ſecundæ diametro A B, utcunque ductæ in Q: dico, ſi ex C ad eandem diametrum A B ordinatim applicetur recta C B, & ex A eidem æquidiſtans ducatur A K H, ſecans contingentem F C Q in I, Hyperbolæque occurrens in K, atque per K recta agatur D K G ipſi A B parallela, (ita ut [1] A K H diameter ſit ſecundæ diametro A B conjugata, ac ſemi-ſecundæ diametri magnitudine ſint K G, K D,) fore rectangulum B A Q æquale ipſius K G vel K D ſemi-ſecundæ diametri quadrato.

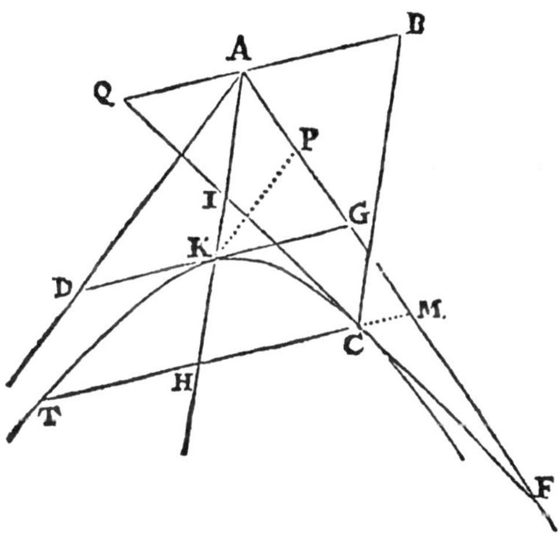

[1] *per 7 hujus.*

Ducta enim per C recta T C M ſecundæ diametro A B parallela, ideoque ad interceptam diametrum A K H ordinatim applicata, quæ Hyperbolæ occurrat in T diametroque A H in H, Aſymptoto verò A F in M: Quoniam eſt [2] H A quadratum ad K A quadratum, ſive [3] H M quadratum ad K G quadratum ſeu

[2] *per 11 hujus, & Cor. 20 ſexti.*
[3] *per 4 & 22 ſexti.*

[201] (or[1] to the rectangle TMC) as HA or CB is to IA, i.e.,[2] as BQ is to AQ, therefore–by separation[3]–the square on HC or the square on BA will be to the square on KG as BA is to AQ. Therefore[4] BA, KG and AQ will be continuedly proportional and the rectangle BAQ[5] will be equal to the square on KG. Which was to be proved.

Corollary to the two preceding propositions

From the preceding it is very easy to infer how to draw from an arbitrary given point a straight line that touches a given hyperbola [2.32].

Indeed, if the given point lies on the curve itself, e.g., K, then we first construct the asymptotes[6] and draw to one of them a straight line parallel to the other asymptote, e.g., KP, and choose PG, equal to AP. Then the straight line connecting G, K, and D will touch the hyperbola at K,[7] because GK equals KD[8] as GP equals PA [2.33].

We proceed in the same way if the given point lies on one of the two asymptotes, e.g., G. First we bisect AG at P and draw PK parallel to the other asymptote, meeting the curve[9] at K; then the straight line connecting G, K, and D[10] will touch the hyperbola at the meeting point K.

Next suppose that the given point lies inside the angle enclosed by the asymptotes, e.g., I; then we draw first from the center[11] a diameter through I, e.g., AIH, meeting the curve at K and we take AH as the third proportional to AI and AK; if we draw through H the straight line HC, ordinate-wise applied (which is–as we know–parallel to the tangent at K[12]), meeting the curve at C,[13] then the straight line connecting I and C will touch the hyperbola at this point C [2.34].

Suppose finally that the given point lies inside one of the two angles adjacent to the angle containing the hyperbola, e.g., Q; then we draw through Q and the center A the second diameter QAB and its conjugate transverse diameter AKH (which as we know, being produced, bisects any line segment drawn inside the hyperbola parallel to QAB). We also draw the tangent KG or KD, bounded by an asymptote. If we equalize the rectangle QAB to the square on KG or on KD and if we apply through B the straight line BC ordinate-wise to the second diameter AB (the text gives erroneously AH, transl.), parallel of course to AK,[14] meeting the curve at C, then the line connecting Q and C[15] will touch the hyperbola at this point. Moreover it is clear that, if the given point lies inside the hyperbola or inside the angle opposite to the angle containing the hyperbola, it is not[16] possible to draw from this point a straight line that, being produced, does not intersect this curve [2.36].

---

[1]Prop. 6, Cor. 1, p. [190]; [2]VI, 4; [3] V, 17; [4]VI, 20, Cor.; [5]VI, 17; [6]Prop. 10, Cor. 1, p. [197]; [7]Prop. 6, p. [190]; [8]VI, 2; [9] Prop. 3, Cor. 2, p. [183]; [10]VI, 2 and Prop. 6, p. [190]; [11] that has been found by Prop. 5, Cor. 7, p. [189]; [12]Prop. 6, Cor. 3, p. [191]; [13]Prop. 11, p. [199]; [14]Prop. 7, p. [192]; [15]Prop. 12, p. [200]; [16]Prop. 3, Cor. 1, p. [182].

ſeu[1] ad T M C rectangulum, ut H A ſeu C B ad I A, id eſt[2], ut B Q ad A Q; erit dividendo[3] H C quadratum ſeu B A quadratum ad K G quadratum, ut B A ad A Q. Ac propterea[4] B A, K G, & A Q proportionales erunt, rectangulumque B A Q[5] quadrato K G æquale. Quod demonſtrandum erat.

Corollarium ad duas propoſitiones præcedentes.

Ex dictis facillimè colligitur, quo pacto à dato quolibet puncto ducenda ſit recta, quæ datam Hyperbolam contingat.

Si enim datum punctum in ipſa curva ſit, veluti K, inventis Aſymptotis[6], ductâque ad illarum alterutram rectâ alteri Aſymptoto parallelâ, ut K P, ac ſumptâ P G ipſi A P æquali, continget juncta G K D Hyperbolam in K[7]. quoniam uti G P ipſi P A, ita G K ipſi K D æqualis eſt[8].

Eodem modo, ſi datum punctum ſit in Aſymptotorum alterutra, veluti G, diviſâ A G bifariam in P, ductâque P K alteri Aſymptoto parallelâ, quæ curvæ occurrat[9] in K: continget juncta G K D[10] Hyperbolam in puncto occurſus K.

Sit deinde datum punctum intra angulum Aſymptotis comprehenſum, veluti I: ductâ à centro[11] per I diametro, ut A I H, quæ curvæ occurrat in K, ſumptâque A H ipſis A I, A K tertiâ proportionali, ſi per H agatur ordinatim applicata H C (nimirum, quæ contingenti in K æquidiſtet[12]), occurrens curvæ in C, continget juncta I C[13] Hyperbolam in eodem C puncto.

Sit denique datum punctum in alterutro angulorum, qui deinceps ſunt, angulo Hyperbolam continenti, veluti Q: ductâ per Q & centrum A ſecundâ diametro Q A B, tranſverſâque ipſi conjugatâ A K H (nimirum, quæ producta quamlibet rectam in Hyperbola ductam ipſi Q A B æquidiſtantem bifariam dividat), nec non tangente K G vel K D, Aſymptoto terminatâ; ſi fiat quadrato K G vel K D æquale rectangulum Q A B, ac per B ad ſecundam diametrum A H applicetur recta B C, nempe ipſi A K æquidiſtans[14], quæ curvæ occurrat in C: juncta Q C[15] in eodem puncto C Hyperbolam continget.

Manifeſtum porrò eſt, ſi datum punctum vel intra Hyperbolam foret, vel intra angulum ad verticem ei, qui Hyperbolam continet: fieri non poſſe[16], ut ab eodem puncto ducatur recta, quæ producta eandem non ſecet.

[1] *per 1 Cor. 6 hujus.*
[2] *per 4 ſexti.*
[3] *per 17 quinti.*
[4] *per Cor. 20 ſexti.*
[5] *per 17 ſexti.*
[6] *per 1 Cor. 10 hujus.*
[7] *per 6 hujus.*
[8] *per 2 ſexti.*
[9] *per 2 Cor. 3 hujus.*
[10] *per 2 ſexti. & 6 hujus.*
[11] *invento per 7 Corol. 5 hujus.*
[12] *per 3 Cor. 6 hujus.*
[13] *per 11 hujus.*
[14] *per 7 hujus.*
[15] *per 12 hujus.*
[16] *juxta 1 Cor. 3 hujus.*

[202]

Chapter III
Third Definitions

I

Suppose that an arbitrary side of a right-angled triangle –whether it subtends the right angle or is opposite to one of the acute angles– moves within this angle in such a way that each endpoint of this moving side always lies and remains on the side with which it was initially connected, and suppose that this side, if necessary, has been produced in one direction or in both directions.

Suppose also that this motion is continued successively through the adjacent angles of the first mentioned angle, as well as through the opposite angle, until the moving side has returned to its original position and place, and assume that in this way by an arbitrary point that has been marked on this segment or –if desired– even on its produced part, a curve is described, then the said moving side will be indicated by the name the generator [3.1].

II

The point, however, that has been marked on this straight line to describe the curve will be called the active point or simply the point.

III

The distance from this point to one of the endpoints of the generator, as well as to the other endpoint, will be called the interval.

IV

When we refer simply to the angle, then we mean that angle that is subtended by the generator and within which the generator moves.

V

The vertex of the angle, about which the generator, as it were, moves around in a continuous motion, will be called the center.

CAPUT III.

DEFINITIONES TERTIÆ.

I.

SI quodlibet trianguli rectanguli latus, sive id rectum angulum subtendat, sive acutorum alterutri opposi-tum sit, in eodem angulo moveatur, ita ut uterque mo-ti lateris terminus semper existat, maneatque in latere, cui ab initio junctus fuit, producto tamen sive ab altera sive ab utraque parte, prout opùs fuerit; idemque ille motus tam per angulos, qui præfato deinceps sunt, quàm per eum, qui ipsi ad verticem est, ordine con-tinuetur, donec ad positionem situmque pristinum latus motum redierit, atque ita quolibet puncto quod in eodem, utcunque etiam producto, notare placue-rit, curva describatur linea, prædictum mobile latus *Describentis Lineæ* nomine designabitur.

II.

Punctum autem quod in eodem ad descriptionem notare placuerit, *Punctum Efficiens*, aut *Punctum* sim-pliciter vocabitur.

III.

Distantia verò ejusdem puncti tam ab uno quàm ab al-tero *describentis* termino *Intervallum* dicetur.

IV.

Cum de *angulo* simpliciter sermo erit, eum intellige-mus, quem subtendit, & in quo movetur *describens*.

V.

Anguli vertex, quem *describens* continuato motu quasi circumambulat, *Centrum* appellabitur.

VI.

[203]

VI

If one measures along one of both arms of the angle (produced, if necessary, on both sides) from the center, on both sides, the length of the interval terminated by the other arm, then this –total– part of the arm will be called the directrix.

VII

We will say that the generator is in the initial position when it is perpendicular to the directrix. The same, however, will then also be said about the point, and when they are referred to in this way, then we assume that they are in this position.

VIII

Twice the part of the line segment drawn from the point (in the initial position, transl.) through the center, intercepted between the point and the center, will be called the secant.

Suppose that the side BC of the right-angled triangle ABC moves inside the angle BAC, e.g., in such a way that the endpoint C approaches A and that at the same time B recedes from I or moves forward to I, in such a way, however, that the endpoints B and C always lie and exactly remain on the sides with which they were initially connected, namely B on the side AB and C on the side AC, both produced if necessary.

Suppose moreover that this side (BC, transl.) describes a curve by means of one of its points, e.g., H. This point has been chosen arbitrarily, be it on BC itself or on the produced part of BC (as we suppose usually, because it seems in a sense better suited). This description runs as follows:

when the point C has arrived at A and the point B at I and when at the same time H has proceeded to F,

VI.

Alterutrum *anguli* crus, utrinque, si opùs fuerit, productum, atque ab utraque parte à *Centro* sumptum, magnitudine *intervalli* in altero crure terminati *Directrix* vocabitur.

VII.

Describentem in statione prima dicemus, cùm ea ad *directricem* est perpendicularis: idem autem & tunc de *puncto* dictum esto, ac cùm de iis simpliciter sermo erit in ea statione considerabuntur.

VIII.

Recta à *puncto* per *Centrum* ducta, interceptæ inter *punctum* & *centrum* dupla, *Secans* nuncupabitur.

Ut si trianguli rectanguli A B C latus B C moveatur in *angulo* B A C, ex. gr., ut terminus C tendat ad A, simulque B vel retrocedat vel promoveatur versùs I; ita tamen, ut iidem termini B & C semper sint & exactè maneant in lateribus, quibus ab initio juncti fuère, nempe B in latere A B, ac C in latere A C, productis ubi opùs fuerit; eodemque illo motu quolibet sui puncto. ex. gr., H, assumpto, prout placuerit, sive in ipsa B C, sive in eadem producta, (ut à nobis plerumque assumetur, cum id naturæ quodammodo convenientiùs videatur,) describat curvam lineam: nempe, ut, ubi punctum C pervenerit ad A, ac punctum B ad I, simulque H processerit ad F, descripta sit per motum puncti

Fig. 1.

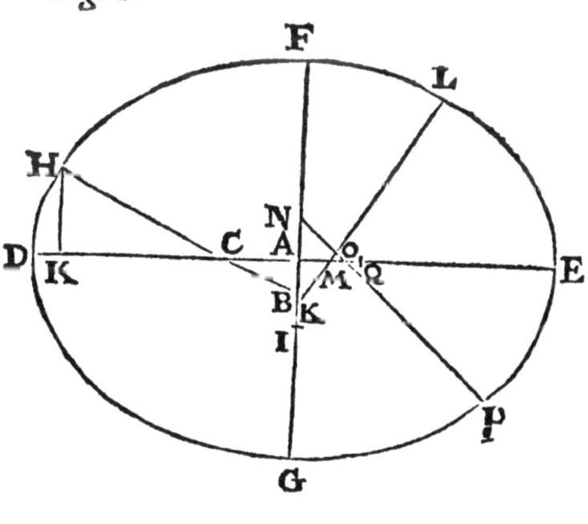

[204] then by the motion of the point H, the part HF of the curve has been described. Suppose next that the arc FL has been described by the motion of the point C via A to M, while at the same time B recedes or proceeds from I to K in such a way that H arrives at L, and that in the same way the arc LE is described when the point B has arrived at A by a continuous motion via K, while at the same time the point C, proceeding via M, has arrived at Q and the point H coincides with E.

Finally, let the arc EP be described when the point B has proceeded via A to N and at the same time the point C has receded or proceeded from Q to O in such a way that the point H has then arrived at P.

If next the motion is continued in the same way until the said point has passed via G and D and has again reached the point H, then the whole curve $HFLRPGD$ has been described and the following will be true:

> BC–which in other positions is IA, KM, AQ, NO, etc.– is the generator.
>
> H is the active point.
>
> HC and HB are both intervals.
>
> The vertex of the angle, namely the point A, is the center.

Suppose that one of both arms of the angle, e.g., AC, has been produced on both sides, if necessary, e.g., to D and E, in such a way that AD as well as AE is equal to the line segment HB –which is, as we know, the interval that is bounded by the other arm– then the whole DE will be the directrix.

Next we consider the case in which the generator BC is perpendicular to this directrix DE. This happens when it is in the same position as the arm AB, e.g., AI, if the angle is a right angle (as in the first figure) or –if the angle

H curvæ portio HF: deinde puncto C promoto per A ad M, simulque termino B retrogresso vel progresso ab I ad K, ita ut H pervenerit ad L, descriptus sit arcus FL: eodemque modo, ubi punctum B per K continuato motu pervenerit ad A, simulque punctum C per M progrediendo pervenerit ad Q, ac punctum H in E inciderit, descriptus sit arcus LE: ac rursus ubi punctum B per A progressum fuerit ad N, simulque punctum C ex Q vel retrocesserit vel progressum sit ad O, ita ut tunc punctum H pervenerit ad P, descriptus sit arcus EP: atque si porrò eodem pacto motus ille continuetur, donec prædictum punctum per G & D transierit rursusque ad H pervenerit, descripta sit tota curva HFLEPGD: erunt

Fig. II.

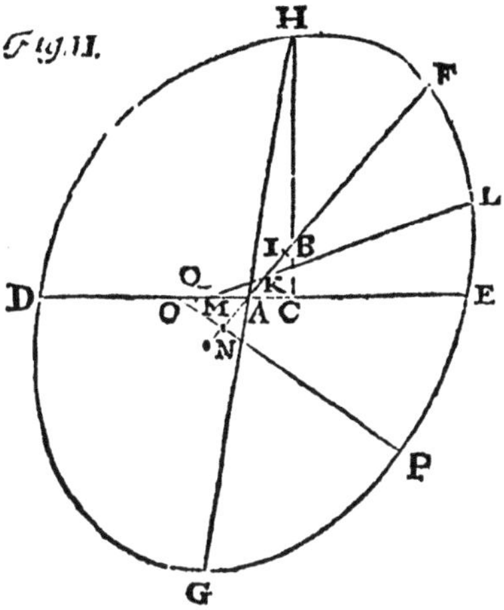

BC, quæ & in aliis stationibus est IA, KM, AQ, NO, &c. *linea describens.*
H *punctum efficiens.*
HC & HB utrumque *intervallum.*
Anguli vertex, nempe punctum A, *Centrum.*

Fig. III.

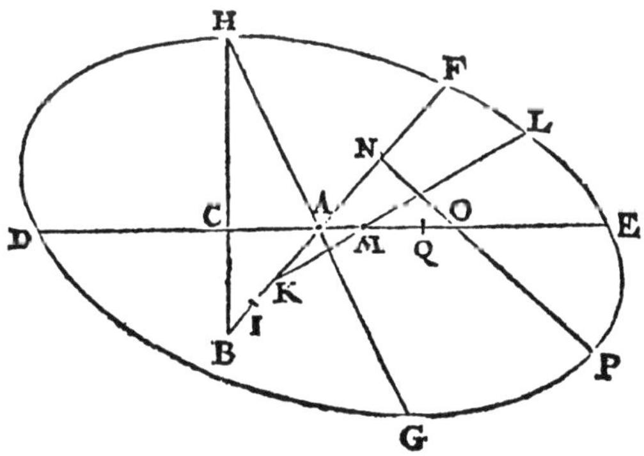

Et si alterutrum *anguli* crus, exempli gratiâ, AC, utrinque, si opùs fuerit, productum sit, veluti ad D & E; ita nempe, ut tam AD quàm AE æqualis sit rectæ HB, *intervallo* videlicet, quod in altero crure terminatur, tota DE *directrix* erit.

Cum autem *describens* BC eidem *directrici* DE est perpendicularis, quod quidem fit, quando ipsa positione eadem est cum crure AB, uti AI, *angulo* nempe existente recto, ut in

[205] is oblique– when it is in the same position as BC, as is shown in the following figures. In the first case IA, and in the second case BC, will be the generator in the initial position, or simply the generator. Therefore the point F or H, which lies in the same direction, will also be called the active point in the initial position, or simply the point. Likewise FAG[a] or HAG[b], drawn namely from the point through the center and twice FA or HA, represents the secant [3.2].

Theorem XII

Proposition 13

A curve, described according to the definitions set forth in this chapter, in an arbitrary angle and with arbitrary intervals, has the property that the square on an arbitrary line segment applied to the curve from an arbitrary point on the directrix and parallel to the secant, has the same ratio to the rectangle contained by the parts of the directrix into which it (the directrix, transl.) is divided by the applied line segment, as the square on the secant has to the square on the directrix [3.3].

Suppose that in an arbitrary angle BAC the curve $DHEG$ has been described with arbitrary intervals HC and HB and with directrix DAE and secant FAG[a] or HAG[b] [3.4]. Suppose moreover that from the point I – arbitrarily chosen on the directrix DE – the line segment IL has been applied to the curve, parallel to the secant FAG[a] or HAG.[b] I say that the square on the applied LI will be to the rectangle DIE as the square on the secant FG[a] or HG[b] is to the square on the directrix DE.

Indeed, let the straight line KM be the generator in the position in which it was when the point L was described by it. First we consider the case in which the angle BAC is a right angle[a] [3.5] and we draw KN, parallel to the directrix DE, intersecting the applied segment LI, produced if necessary, at N. Then the following is true: the interval KL equals half the directrix i.e., AE or AD, and so the square on KL is equal to the square on AE or AD too; if one subtracts equals on both sides, namely[1] the square on KN on one side and the square on AI on the other, then the remainders, namely the square on LN and the rectangle DIE[2] will be equal too. However,[3] the square on LI is to the square on LN, i.e.,[4] to the rectangle DIE,

---

[a]in the case of Fig. I and the like; [b]in the cases of Figs II and III and the like; [a]in the cases of Figs. I, II and the like; [b]in the cases of the other and similar figures; [1]I, 34; [2]I, 47 and II, 5; [3]VI, 4 and 22; [4]by what has been proved above.

ut in prima figura, aut si obliquus fuerit *angulus*, in ipsa positione B C, uti exhibetur in sequentibus figuris, erit I A, casu primo, & B C, casu altero, *describens in statione prima* seu *describens* simpliciter, ideoque punctum F vel H, quod eidem in directum est, *punctum efficiens in statione prima* seu *punctum* simpliciter.

Ac proinde F A G [a] vel H A G [b], nempe ab eodem *puncto* per centrum A ducta atque ipsius F A sive H A dupla, *secantem* repræsentat. [a] in casu fig. I & similib. [b] in casu fig. II & III ac similib.

THEOREMA XII.

Propositio 13.

In quocunque *angulo*, & quibuslibet *intervallis*, juxta definitiones hoc capite propositas, curvâ descriptâ, hoc ipsi proprium erit, ut quadratum cujuslibet *secanti* æquidistantis, à quolibet *directricis* puncto ad curvam applicatæ, eandem rationem habeat ad rectangulum sub partibus *directricis* per applicatam factis, quam quadratum *secantis* ad quadratum *directricis*.

Sit in quocunque *angulo* B A C, *intervallis* quibuslibet H C, H B, descripta curva D H E G, cujus *directrix* D A E, *secans* [a] F A G [a] vel H A G [b]; atque à puncto I in *directrice* D E utcunque assumpto, ad curvam applicata I L *secanti* F A G [a] vel H A G [b] æquidistans: dico fore quadratum applicatæ L I ad rectangulum D I E, ut est quadratum *Secantis* F G [a] vel H G [b] ad quadratum *directricis* D E. [a] in casib. fig. I, II, & similibus. [b] in casib. cæterarum fig. & similibus.

Sit enim recta K M *describens* in ea statione, uti fuit, cùm per eandem descriptum est punctum L. Et primò quidem, si *angulus* B A C rectus sit [1], ductâ K N *directrici* D E parallelâ, quæ occurrat applicatæ L I, aut eidem productæ, si opùs fuerit, in N: cum *intervallum* K L æquale sit dimidiæ *directrici* A E vel A D, ideoque & K L quadratum æquale A E vel A D quadrato, ablatis utrinque æqualibus, nimirum [2] quadrato K N ab una, & quadrato A I ab altera parte, residua quoque, nempe L N quadratum & D I E rectangulum [2], æqualia erunt. Unde cum [3] sit ut L I quadratum ad L N quadratum, id est [4], ad D I E rectan- [1] per 34 primi. [2] per 47 primi, & 5 secundi. [3] per 4 & 22 sexti. [4] per supra demonstr.

 gulum,

[206] as the square on LM is to the square on LK, i.e., as the square on FA is to the square on AE or[1] as the square on FG is to the square on DE.

Therefore the proposition proves to be right in the first case. Secondly, let the angle BAC not be a right angle[b] [3.6] and suppose that the straight lines KO and LP are drawn to the directrix, produced if necessary, parallel to the generator BC (in the initial position, transl.) and thus perpendicular to the directrix DE, and let also be drawn IN parallel to the arm AB, intersecting LP -produced if necessary- at N, so that[2] the triangles AHC and ILP are similar, as well as AHB and ILN.

---

[1]V, 15; [2]I, 29.

Fig. I.

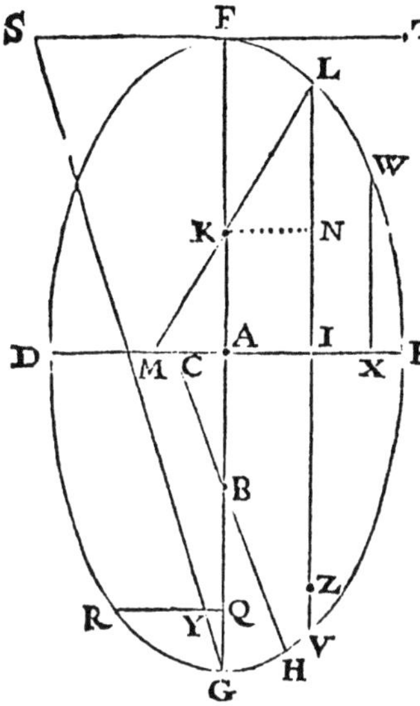

Fig. II.

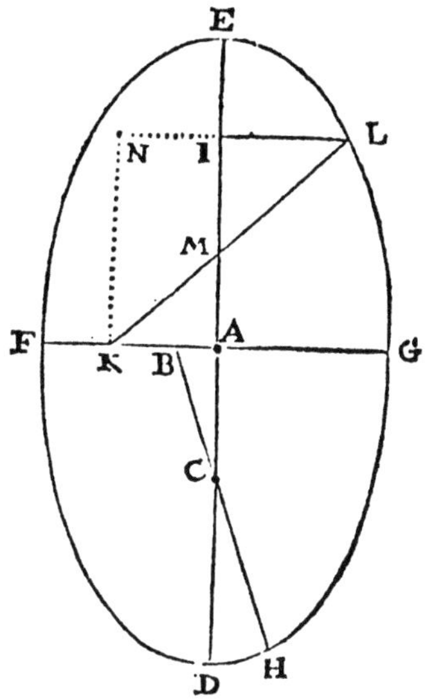

1 per 15 quinti. gulum, ita LM quadratum ad LK quadratum, hoc est, ita FA quadratum ad AE quadratum, sive [1] ut FG quadratum ad DE quadratum, constat priori casu propositum.

Fig. III.

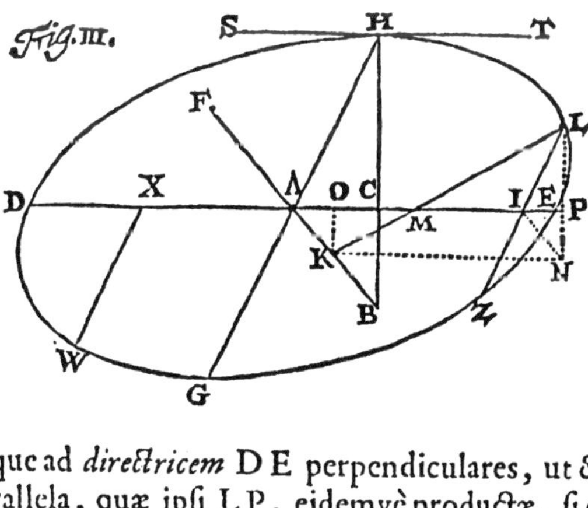

Non sit deinde *angulus* BAC rectus[b], ducanturque ad *directricem*, eamve productam, si opus fuerit, rectæ KO, LP *describenti* BC parallelæ, ideoque ad *directricem* DE perpendiculares, ut & IN lateri AB parallela, quæ ipsi LP, eidemve productæ, si opus fuerit, occurrat in N; ita ut [2] similia sint triangula AHC & ILP, itemque

2 per 29 primi.

[207] Let us finally connect K and N.

Now we reason as follows: BA is to KA,[1] or BC, i.e., MK, is to KO as ML, i.e., HC, is to LP. However, HC is to LP as HA is to LI and as BA is to NI, so consequently BA is to KA as the same BA is to NI; therefore[2] KA will be equal to NI. They are, however, parallel too, by hypothesis. Therefore AI and KN will also be equal and parallel.[3]

Moreover, the line segments KL and EA or AD are equal and so, therefore, are their squares; if we subtract from them equals, namely the square on KN on one side and the square on AI on the other side, then the

[1] VI, 4; [2] V, 9; [3] I, 33.

Fig. IV.

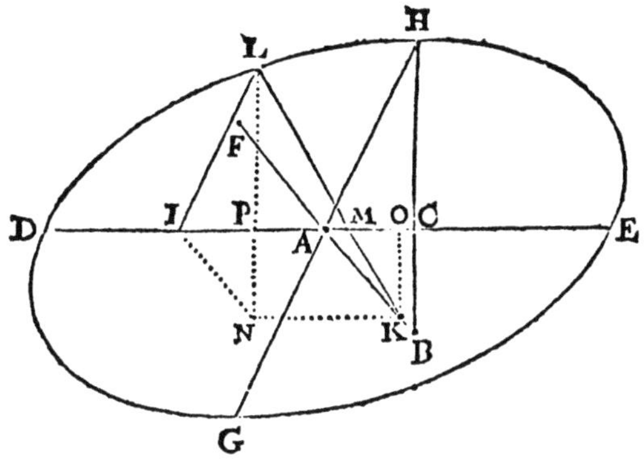

Fig. V.

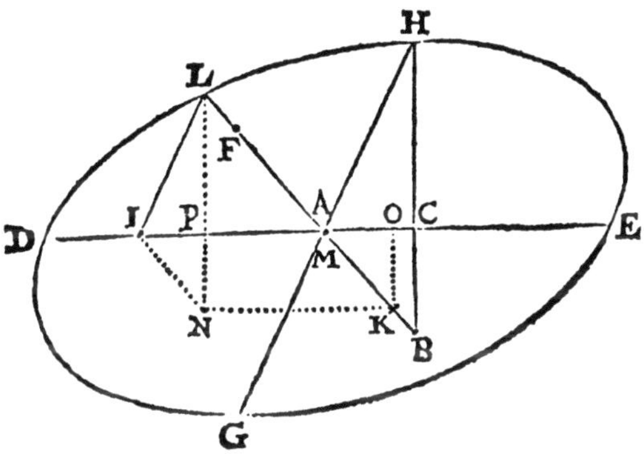

Fig. VI.

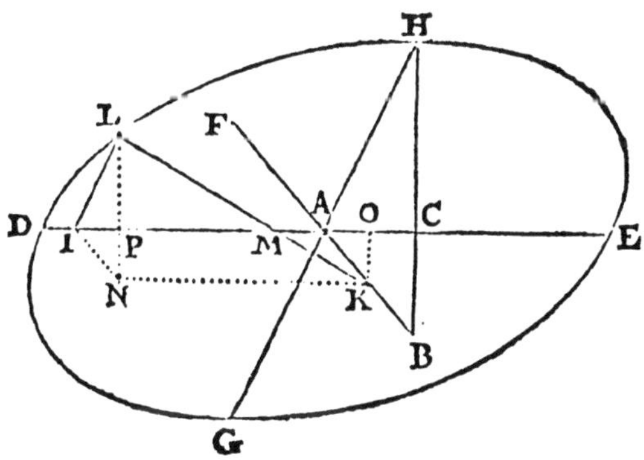

que A H B & I L N, ac denique jungatur K N. Quoniam itaque est [1], ut B A ad K A, sive ut B C, id est, M K, ad K O, ita M L, hoc est, H C, ad L P; ut autem H C ad L P, ita H A ad L I, & ita B A ad N I, ac per consequens B A ad K A, ut eadem B A ad N I: erit [2] K A ipsi N I æqualis. Sunt autem & parallelæ, ex hypothesi. Quare & A I, K N æquales & parallelæ erunt [3]. Porrò cum æquales sint rectæ K L & A E vel A D, ideoque & ipsarum quadrata, hinc subductis ab iis æqualibus, quadrato nimirum K N ab una, ac quadrato A I ab altera parte, erunt

[1] per 4 sexti.

[2] per 9 quinti.

[3] per 33 primi.

[208] remainders, namely the square on LN and the rectangle DIE, will be equal too.[1] From this we conclude: because[2] the square on LI is to the square on LN, i.e.,[3] to the rectangle DIE, as the square on AH is to the square on HB, i.e., to the square on AE, or[4] as the square on HG is to the square on DE, therefore in this case too the proposition will be clear.

Thus it is clear that the said curve is the same as the curve that was called "ellipse" by the Ancients and that the directrix and secant are the same as what they called conjugate diameters or –if the angle should be a right angle– conjugate axes.

Therefore we will call two line segments drawn through the center and bounded by the ellipse on both sides, conjugate diameters

---

[1] I, 47 and II, 5; [2] VI, 4 and 22; [3] by what has been proved above; [4] V, 15.

Fig. VII.

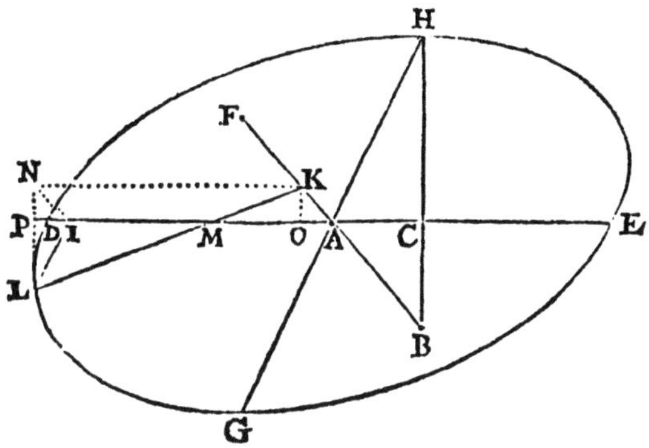

Fig. VIII.

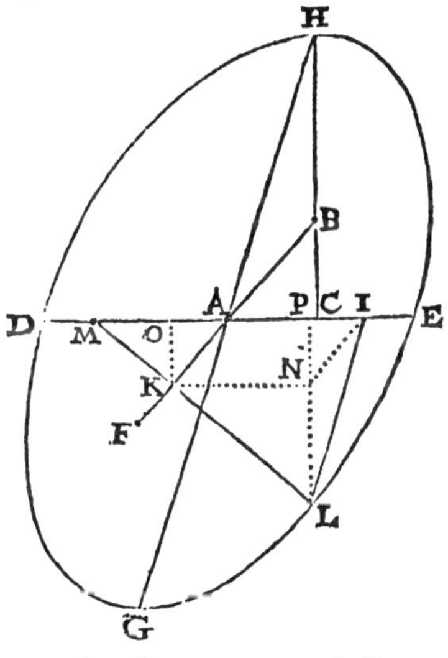

erunt quoque residua, quadratum nempe L N & rectangulum D I E æqualia [1]. Unde cum sit [2] L I quadratum ad L N quadratum, hoc est [3], ad D I E rectangulum, ut A H quadratum ad H B quadratum, id est, ad A E quadratum, sive,[4] ut H G quadratum ad D E quadratum, erit etiam hoc casu propositum manifestum.

[1] *per 47 primi, & 5 secundi.*
[2] *per 4 & 22 sexti.*
[3] *per supra demonstr.*
[4] *per 15 quinti.*

Atque ita liquet, prædictam curvam eam ipsam esse, quæ Veteribus Ellipsis dicta fuit, *directricem* verò ac *secantem* eas ipsas, quas conjugatas diametros, aut, si *angulus* rectus fuerit, conjugatos axes vocârunt.

Conjugatas itaque diametros appellabimus binas rectas per centrum ductas, ac utrinque Ellipsi terminatas;

[209] if they have the following property (as has already been proved in the case of the directrix and the secant): the squares on the segments, applied to one of them (viz. to one of the conjugate diameters, transl.) and parallel to the other, have to the rectangle contained by the parts cut off by the applied segment, the same ratio as the square on one of them has to the square on the diameter that is intersected by the applied segment [3.7].

The diameter on which the applied lines rest will be called the transverse diameter, but the diameter to which they are parallel will be called the second diameter.

All other line segments, however, drawn through the center and bounded by the ellipse on both sides, will simply be called diameters.

The line segment that is the third proportional to the transverse and the second diameter will be called the latus rectum or the parameter related to the transverse diameter [3.8].

It must be noticed, however, that –if the angle is a right angle and the point has the same distance to each of the endpoints of the generator– the curve that is described by the point in the above-mentioned way is the circumference of a circle.

Corollary 1

The following is clear from the demonstration and from the comparison of the first figure with the second: the transverse axis of two conjugate axes in an ellipse is also second axis and vice versa. Indeed, whether LI has been applied to one axis or to the other, one can always prove –in the same way– that the square on this applied segment is to the rectangle contained by the parts of the axis to which it is applied, as the square on the other axis is to the square on the said axis that is intersected by the applied line segment [3.9].

Corollary 2

Moreover, it is clear that the straight line drawn through the point (in the initial position, transl.) parallel to the directrix, i.e., the straight line that

tas; ita ut (quemadmodum de *directrice* & *secante* jam demonstratum est,) quadrata rectarum quæ alteri ipsarum applicantur alteri æquidistant, ita se habeant ad rectangula sub partibus per applicationem factis, ut quadratum alterius ad quadratum ejusdem quæ per applicatas secatur.

Et hæc quidem, cui applicatæ insistunt, ttansversa; illa verò, cui eædem æquidistant, secunda diameter vocabitur.

Cæteræ autem omnes, per centrum ductæ ac utrinque Ellipsi terminatæ, diametri simpliciter dicentur.

Rectam lineam quæ transversæ secundæque diametro tertia est proportionalis, Latus Rectum sive Parametrum vocabimus ad transversam diametrum pertinentem.

Notandum tamen est, si *angulus* rectus sit, ac *punctum* ab utroque *describentis* termino æqualiter distet, curvam, quæ motu ejusdem *puncti*, uti prædictum est, describitur, circumferentiam Circuli esse.

Corollarium. I.

Ex ipsa demonstratione & collatione figuræ primæ cum secunda manifestum est: in Ellipsi, conjugatorum axium transversum etiam secundum esse, & contra. Sive enim LI vel huic vel illi axi applicata sit, eodem modo semper probabitur esse quadratum ejusdem applicatæ ad rectangulum sub partibus axis cui applicatio fit, ut quadratum axis alterius ad quadratum axis prædicti qui per applicatam secatur.

Corollarium 2.

Apparet porrò rectam per punctum ductam *directrici* parallelam, hoc est, eam, quæ per terminum secundæ diametri trans-

[210] is drawn through the endpoint of the second diameter, parallel to the transverse diameter, touches the ellipse at this point and at no other point. It would certainly not intersect the ellipse.

Indeed, if one draws through F[a] or H[b] (the endpoint of the second diameter GF[a] or GH[b]) the straight line ST, parallel to the diameter DE, and if one chooses another arbitrary point on the curve, e.g., L, that was generated when the generator took the position KM, and if one draws

[a]in the case of Fig. I and the like; [b]in the case of Fig. III and the like

verſæ æquidiſtans ducitur, Ellipſin in eodem termino, & in nullo præterea puncto contingere, multò minus eandem ſeca-

Fig. I.

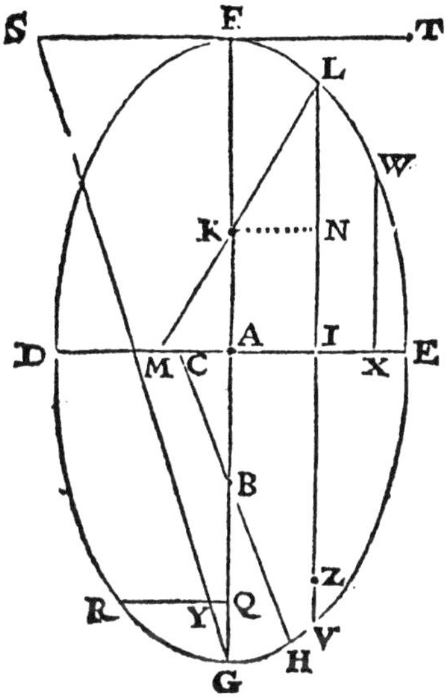

Fig. II.

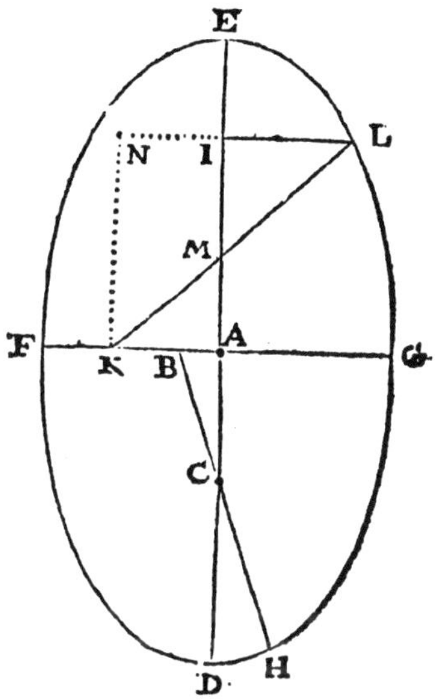

re. Si enim per F[a] aut H[b] terminum ſecundæ diametri G F[a] vel G H[b] ductâ rectâ S T, tranſverſæ diametro D E paralle-

[a] in caſu fig. I & ſimilib.
[b] in caſu fig. III & ſimilib.

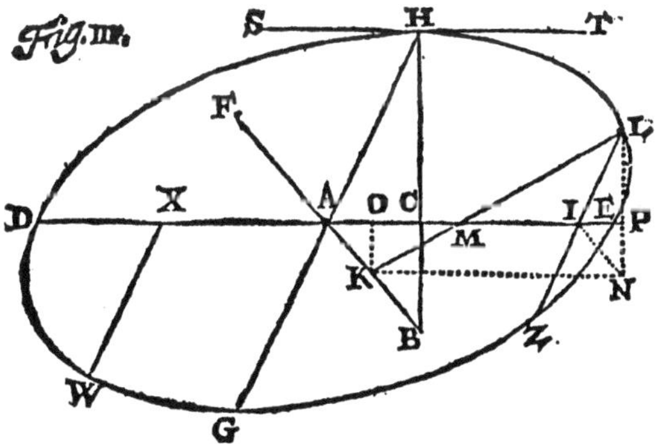

lâ, aſſumatur aliud quodcunque in curva punctum, veluti L, quod deſcriptum ſit *deſcribente* in ſtatione K M, ducaturque

L I

[211] LI[a] or LP[b] perpendicular to the transverse diameter, then it will happen that in the triangle MLI[a] or MLP[b] the line segment ML, i.e., the perpendicular FA[a] or HC,[b] is greater[1] than LI[a] or LP,[b] so that the point L, arbitrarily chosen on the curve –and so the whole curve, with the exception of F[a] or H[b]– lies below the drawn ST or nearer to the center.

Corollary 3

It is also clear that in an ellipse the squares on ordinate-wise applied segments are to each other as are the rectangles contained by the parts of the diameter, cut off by the applied segments. E.g., if LI and WX have been applied, then the square on WX will be to the rectangle DXE as the square on LI is to the rectangle DIE, since[2] both ratios are the same as the ratio of the square on FG[a] or HG[b] to the square on DE, or as the ratio of the parameter to the transverse diameter, and so–by interchanging the means–as the ratio of the rectangle DXE to the rectangle DIE [3.10].

Corollary 4

It is also clear that line segments that are ordinate-wise applied to an axis or to a diameter are bisected by this axis or diameter, if they have been produced to the ellipse. E.g., if the applied segment LI, being produced, intersects the ellipse at V, then the square on LI[3] is to the rectangle DIE as the square on VI is to the same rectangle DIE; therefore[4] the square on LI will be equal to the square on VI and so the segment LI itself will be equal to the segment VI [3.11].

Corollary 5

Furthermore it is clear that ordinate-wise applied lines cannot intersect an ellipse at more than two points. Indeed, if LIV should have –distinct from L and V– another point, e.g., Z, on the ellipse, then the segments IL and IZ[5] and so also IV and IZ, would be equal as part and total (of the same, transl.), which is absurd [3.12].

Corollary 6

From the preceding, one can moreover gather the following: if one draws

---

[a]in the case of Fig. I and the like; [b]in the case of Fig. III and the like; [1]I, 18; [2]Prop. 13, p. [205]; [3]by the preceding Cor.; [4]V, 9; [5]by the preceding corollary .

LI [a] vel LP [b] ad transversam diametrum perpendicularis, fiet ut in triangulo MLI [a] vel MLP [b] recta ML, id est, perpendicularis FA [a] vel HC [b], major sit [1] quàm LI [a] vel LP [b]; adeò ut punctum L, quod in curva utcunque assumptum est, id est, tota Ellipsis, præter F [a] aut H [b] punctum, infra ductam ST, seu versùs Ellipseos centrum, cadat.

[a] in casu fig. I & similib. [b] in casu fig. III & similib. [1] per 18 primi.

Corollarium 3.

Manifestum quoque est in Ellipsi applicatarum quadrata ad se invicem esse, ut rectangula sub diametri portionibus per applicatas factis. Ut si applicatæ sint LI, WX, erit quadratum WX ad rectangulum DXE, ut quadratum LI ad rectangulum DIE: cum [2] utriusque ratio sit eadem quæ quadrati FG [a] vel HG [b] ad quadratum DE, sive quæ parametri ad transversam diametrum; ideoque & permutatim WX quadratum ad LI quadratum, ut DXE rectangulum ad DIE rectangulum.

[2] per 17 hujus.

Corollarium 4.

Constat etiam ordinatim ad axem sive diametrum applicatas utrinque ad Ellipsin productas ab axe sive diametro bifariam secari. Ut, si applicata LI producta Ellipsi occurrat in V, quoniam est [3] quadratum LI ad rectangulum DIE, ut quadratum VI ad idem DIE rectangulum, erit [4] quadratum LI æquale quadrato VI, ideoque & ipsa recta LI ipsi rectæ VI æqualis.

[3] per Cor. præced. [4] per 9 quinti.

Corollarium 5.

Constat porrò, applicatas Ellipsi in pluribus quàm duobus punctis non occurrere. Si enim LIV alio sui puncto præter L & V, exempli gratiâ, puncto Z, in Ellipsi esset, rectæ IL & IZ [5], ideoque IV & IZ pars & totum, æquales forent, quod est absurdum.

[5] per Cor. præced.

Corollarium 6.

Ex dictis porrò colligitur, si ab extremitate transversæ dia-

 metri,

[212] from the endpoint of a transverse diameter, e.g., FG,[a] the parameter FS parallel to the second diameter DE and if one then connects S and G, and if one applies to the same (transverse) diameter an arbitrary line segment, e.g., RQ, intersecting the connecting line SG at Y, then the rectangle FQY will be equal to the square on the applied RQ.

Indeed, because[1] the square on GF is to the square on DE, i.e.,[2] because GF is to FS [3.13] or GQ is to QY, i.e.,[3] because the rectangle GQF

[a]in the case of Fig. I and the like; [1]Prop. 13, p. [205]; [2]VI, 20, Cor.; [3]VI, 1.

[1] in casu fig. I & similib.

metri, ut puta FG[1], eductâ parametro FS secundæ diametro DE parallelâ, jungatur SG, atque ad eandem diametrum recta

Fig. I.

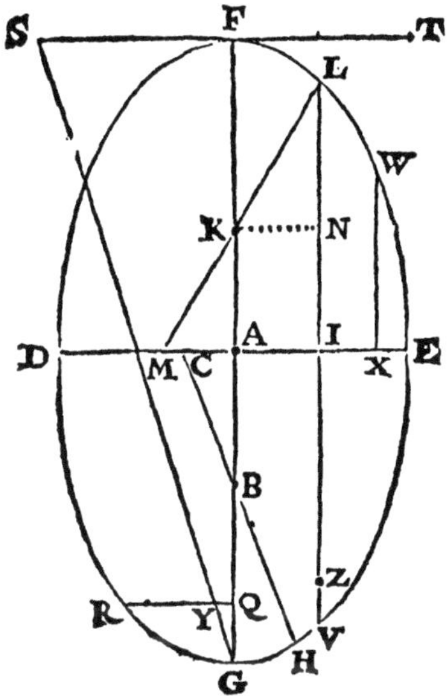

quælibet ordinatim applicetur, ut RQ, quæ secet junctam SG in Y: fore rectangulum FQY quadrato applicatæ RQ æquale.

Fig. III.

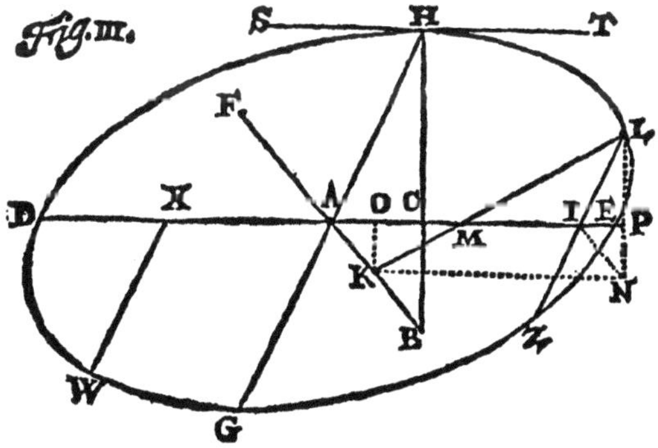

[1] *per 13 hujus.*
[2] *per Cor. 20 sexti.*
[3] *per 1 sexti.*

Quoniam enim est [1] GF quadratum ad DE quadratum, id est [2], GF ad FS, sive GQ ad QY, hoc est [3], GQF rectangulum ad YQF

[213] is to the rectangle YQF, as[1] the same rectangle GQF is to the square on RQ; therefore the rectangle YQF and the square on RQ will be equal. This means –if we may formulate this in the style of the Ancient Geometers: The square on the segment (ordinate-wise) applied from (a point on) an ellipse to a diameter, has the same area as the rectangle adjacent to the latus rectum, having as its breadth the length of the segment that is cut off from the diameter between the applied segment and the vertex of the diameter, diminished by a figure that is similar to, and has the same position as, the figure contained by the transverse diameter and the latus rectum [3.13].

Corollary 7

From what has been said before, it is also clear how to describe an ellipse in the plane if arbitrary conjugate diameters are given. E.g., if the conjugate *axes* DAE and FAG[a] are given and if it is required to describe the ellipse (Figure I, page [212]), then we have to describe a curve with the generator BC, being the difference between the semiaxes AD and AF, but with intervals HC and HB, respectively equal to AF and AD, within the angle DAG; this curve will be the required ellipse. If, however, it is required to describe an ellipse with other, arbitrary conjugate diameters, intersecting at an oblique angle, e.g., DE and HG[b] (Figure III, page [212]), then we proceed as follows: first we drop from the endpoint of one of the conjugate diameters a perpendicular upon the other diameter, e.g., HC; then we measure along this line, produced if necessary, the line segment HB equal to DA or AE, and we draw through B and A the segment BAF [3.14]. If we describe in the angle BAC an ellipse with BC as its generator, but with intervals HC and HB, then this curve will be the required one. So, when a diameter and the (corresponding, transl.) parameter are given, as well as the angle included by this diameter and the straight lines that are ordinate-wise applied to it, then it is at the same time known how to describe an ellipse from this data.

Theorem XIII

Proposition 14

In an ellipse, described on arbitrary axes, any diameter is a transverse diameter and has a conjugate second diameter [3.15].

[1]Prop. 13, p. [205]; [2]V, 9; [a]in the case of Fig. I and the like; [b]in the case of Fig. III and the like.

Y Q F rectangulum, ut [1] idem G Q F rectangulum ad R Q quadratum, æqualia erunt [2] Y Q F rectangulum ad R Q quadratum: id est, si veterum Geometrarum more id proponi placeat. [1] per 13 bujus. [2] per 9 quinti.

Quæ ab Ellipsi ad diametrum applicatur potest spatium adjacens lateri recto, latitudinem habens lineam quæ à diametro inter ipsam applicatam & diametri verticem abscinditur, deficiensque figurâ simili similiterque positâ ei quæ lateribus transverso rectoque continetur.

Corollarium 7.

Patet quoque ex antedictis, quo pacto, datis quibuslibet diametris conjugatis, Ellipsis in plano describatur.

Ut si conjugatis axibus D A E & F A G [a] Ellipsis sit describenda, *describente* B C, quæ semi-axium A D, A F differentia sit, *intervallis* verò H C, H B, ipsis A F, A D utroque utrique æqualibus, in *angulo* D A G, curva describatur, eritque hæc ipsa Ellipsis quæsita. [a] in casu fig. I & similib.

At si aliis quibuslibet conjugatis diametris, obliquè sese intersecantibus, ut D E, H G [b], Ellipsis sit describenda: demissâ à termino unius ad alteram perpendiculari, ut H C, sumptâque in eadem seu in ipsa productâ, si opùs fuerit, rectâ H B ipsi D A vel A E æquali, & per B & A ductâ rectâ B A F, si *describente* B C, *intervallis* verò H C, H B, in *angulo* B A C Ellipsis describatur, erit hæc ea ipsa quæ quæritur. [b] in casu fig. III & similib.

Itaque cum datis diametro parametroque, nec non angulo quem faciunt cum eadem diametro ordinatim ad ipsam applicatæ, conjugatæ quoque diametri datæ sint: simul quoque innotescit, quo pacto & illis datis Ellipsis describatur.

THEOREMA XIII.

Propositio 14.

In Ellipsi circa quoscunque axes descriptâ, ducta quælibet diameter transversa est, habetque secundam sibi conjugatam.

[214] In the ellipse $SYXZ$, with center A but with axes SX and YZ, let an arbitrary diameter DAE be drawn, and let the generator be in the position in which it was when the point D or E was described, so that the intervals are DW and DO.

Then we bring first the generator in reverse position, i.e., in one of the angles adjacent to WAO, e.g., in the position PR, so that the segments AR and AP are equal to AW and AO, in reverse order, namely AR is equal to AO and AP is equal to AW and so the triangle WAO[1] is similar and equal to the triangle PAR; next we have to draw from the point H on the ellipse –that lies on the produced part of PR– the other diameter HAG. I say that the diameter DE is a transverse diameter and that HG is the second diameter conjugate to DE, which means the following: Suppose one first drops HC perpendicular upon DE and measures on this HC, produced if necessary, HB equal to DA. Then, if one draws the straight line BAF through B and A. and next describes an ellipse in the angle BAC with intervals HC and HB, so that in any case DE and HG[2] are conjugate diameters, then I say that this ellipse will be entirely the same as the given ellipse in the sense that one coincides with the other in all points.

Indeed, choose on the given ellipse another arbitrary point L, which has been described by the generator in the position TV, and describe a circle with TV as its diameter; this circle will necessarily also pass through A[3] because the angle TAV is a right angle,[a] and it will also[b] intersect the straight lines BAF and DAE

[1]I, 4; [2]Prop. 13, p. [205] and its Cor. 7; [a]or certainly because either the point T or the point V coincides with the point A, as is the case in Fig. VI; [3]by the converse of III, 31.

Sit in Ellipsi SYXZ, cujus centrum A, axes verò SX, YZ, ducta quælibet diameter DAE; & sic describens OW in ea statione, uti fuit cùm descriptum est punctum D vel E, ita ut intervalla sint DW, DO. Deinde applicatâ eâdem describente in statione reciproca, hoc est, in alterutro angulorum qui ipsi WAO deinceps sunt, veluti PR, ita ut rectæ AR, AP ipsis AW, AO reciprocè sint æquales, nimirum AR ipsi AO, & AP ipsi AW, ac proinde triangulum WAO [1] simile & æquale triangulo PAR, ab Ellipseos puncto H, quod describenti PR in directum est, ducta sit diameter altera HAG. Dico diametrum DE transversam esse, HG autem secundam ipsi DE conjugatam: id est, si, ductâ HC ad DE perpendiculari, in eadem HC, produ-

[1] per 4 primi.

Fig. I.

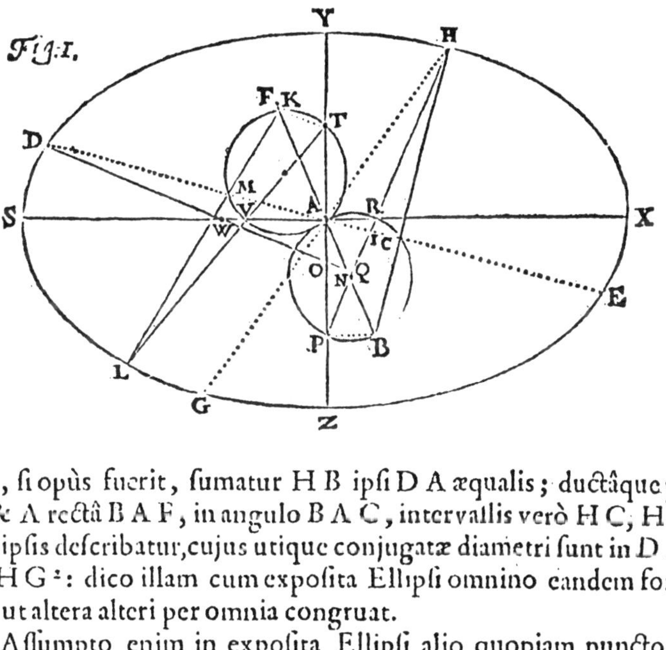

ctâ, si opùs fuerit, sumatur HB ipsi DA æqualis; ductâque per B & A rectâ BAF, in angulo BAC, intervallis verò HC, HB, Ellipsis describatur, cujus utique conjugatæ diametri sunt in DE, & HG [2]: dico illam cum exposita Ellipsi omnino eandem fore, ita ut altera alteri per omnia congruat.

Assumpto enim in exposita Ellipsi alio quopiam puncto L, quod quidem descriptum sit describente in statione TV, diametro TV circulus describatur, qui, propter angulum TAV rectum [4], necessariò quoque per A transibit [3], lineasque BAF & DAE

[2] per 13 hujus ejusque Corol. 7

[4] aut certè, quia punctorum T & V alterutrum cum puncto A coincidit, uti est casus in fig. VI

[3] per conversam 31 tertii.

[215] elsewhere,[b] e.g., in K and M. Next connect K and M, produce KM towards L, and draw TK and PB. Then DO and HP, produced if necessary, intersect at right angles at Q, because the triangle OQP is similar to each of the triangles OAW and RAP[c] and therefore –if we denote the point of intersection of DE and PH by I – the triangles IQD and ICH will be equiangular because the angles at Q and C are right angles and they have the angle at I either in common or as an opposite angle.

Because in the triangles ODA and PHB the sides OD and DA are equal to the sides PH and PB respectively and lie around equal angles, therefore[1]

[b]or will touch one of them and intersect the other, like in the cases of Figs. III and IV; [c]or, if the points O and P coincide, because the angles AOW and APR are each half a right angle; [1]I, 4.

LIB. I. CAP. III. 215

DAE alibi etiam secabit [k], uti in K & M. Deinde junctâ KM, eaque productâ versùs L, agantur TK, PB.

[k] aut illarum alteram continget, alteram verò secabit, ut in casib. fig. III & IV.

Fig. II.

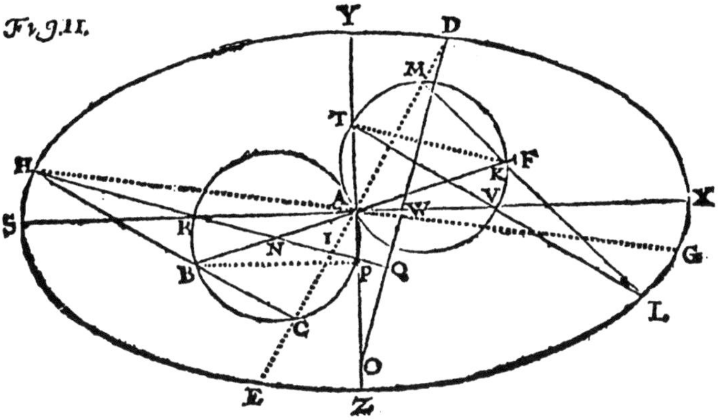

Cum igitur ipsarum DO, HP, productarum, si opùs fuerit, intersectio ad Q fiat ad angulos rectos, ob similitudinem trianguli OQP cum utroque triangulorum OAW, RAP [l], notato ipsarum DE, PH intersectionis puncto I, erunt triangula

[l] vel, si puncta O & P coïncidant, ob angulos AOW, APR semirectos

Fig. III.

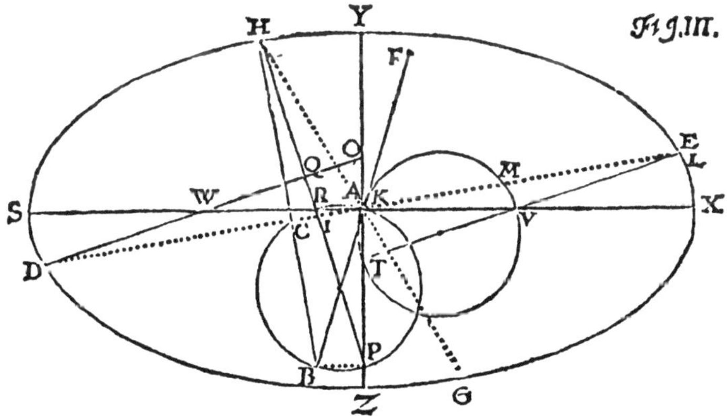

IQD, ICH æquiangula, ob angulos ad Q & C rectos, ad I verò aut communem aut ad verticem. Ideoque cum triangula ODA, PHB latera OD, DA lateribus PH, HB, utrumque utrique, circum æquales angulos æqualia habeant; erit & [m] basis [m] per 4 primi.

OA,

[216] the base OA, or the segment AR, is also equal to the base PB and the angle DOA, i.e., PRA, is equal to the angle HPB, and for that reason[1] the straight line PB is parallel to RA.

In the triangles RAP and BPA the sides RA and AP are respectively equal to the sides BP and PA and lie around equal –namely right– angles, so the base AB will also[2] be equal to the base PR, or to the generator TV, and the angle ARP will be equal to the angle PBA. For this reason the circle, described with diameter AB (which passes through the points C, P,[3] and R,[4] since the angles ACB and APB are right angles and ABP and ARP are congruent), will also be equal to the circle TKV.

The angles PBC and BPR are equal to TKM and KTV, respectively; for obviousy PBC equals TKM,[d] as each of them forms together with the angle PAC or TAM[5] two right angles,[e] and BPR equals KTV because each of them[6] equals the angle BAR or KAV or FAV.[f] Furthermore: because the angles PAB and TAK[7] are equal, so the arcs PB and TK, as well as their chords,

[1] I, 27; [2] I, 4; [3] by the converse of III, 31; [4] by the converse of III, 21; [d] in the case of Fig. II, add: and therefore also their adjacent angles; [5] III, 22; [e] in the case of Fig. II, either is equal to the angle PAC or TAM because of III, 20; in the case of Fig. III, either constitutes two right angles with the angle PAC, one by I, 13 and and the other by III, 22; in the case of Fig. IV, the angles PBC and TKM are equal, because the former constitutes with PAC two right angles and so does the latter with TVA because of III, 22. PAC and TAV, however, are equal because of III, 32. In the case of Fig. V, TKM or TAM is equal to the angle PBC, because either of them constitutes two right angles with the angle PAC because of I, 13 and 20, and III, 22. In the case of Fig. VI, the angles PBC and TKM are equal, because the angle PAC is equal to the inscribed angle on the arc AVM by III, 32; the former of these (scil. PAC, transl.) forms by III, 22 two right angles with the angle PBC, the latter (scil. AVM, transl.), however, forms, with TKM two right angles, also by III, 22; [6] III, 20 and in the case of Fig. III, by III, 20 and III, 32; [f] in the cases of Figs. IV and V, both the angle BPR or BAR and the angle KTV constitute two right angles with KAV by I, 13 and III, 22; [7] III, 26 and 29.

O A, ſive recta A R, baſi P B, angulusque D O A, id eſt P R A,
[1] *per 27 primi.* angulo H P B æqualis; ac propterea [1] recta P B ipſi R A paralle-
la. Hinc cum triangulorum R A P & B P A latera R A, A P la-
teribus B P, P A circa æquales angulos, nempe rectos, utrumque
[2] *per 4 primi.* utrique ſint æqualia: erit & [2] baſis A B baſi P R, ſeu deſcri-
benti T V, angulusque A R P angulo P B A æqualis. Quocirca
& circulus diametro A B deſcriptus (qui quidem, ob angulos
A C B, A P B rectos, ac A B P, A R P æquales, per puncta C,
[3] *per converſam 31 tertii.* P [3], & R [4] tranſit) circulo T K V æqualis erit. Unde cum an-
guli P B C, B P R ipſis T K M, K T V, uterque utrique, æqua-
[4] *per converſam 21 tertii.* les ſint, nempe P B C ipſi T K M [d], quoniam uterque cum an-
gulo P A C ſeu T A M [5] binos rectos conſtituit [e], & B P R ipſi

[d] pro caſu fig. II adde: ideoque & hi qui ipſis deinceps ſunt. [5] *per 22 tertii.* [e] in caſu fig. II uterque angulo P A C ſeu T A M æqualis eſt *per 20 tertii.* In caſu fig. III uterque cum angulo P A C binos rectos conſtituit, nempe hic *per 13 primi*, & ille *per 22 tertii.* In caſu fig. IV æquales ſunt anguli P B C, T K M, quoniam prior cum angulo P A C, poſterior verò cum angulo T V A (qui quidem P A C, T V A æquales ſunt *per 32 tertii*) binos rectos conſtituit *per 22 tertii.* In caſu fig. V, T K M ſive T A M æqualis eſt angulo P B C, quia uterque cum angulo P A C duos rectos conſtituit *per 13 primi & 20 ac 22 tertii.* In caſu fig. VI æquales ſunt anguli P B C, T K M, quoniam angulus P A C æqualis eſt ei, qui in ſegmento A V M conſtitueretur *per 32 tertii*, quorum quidem prior cum angulo P B C, poſterior verò cum angulo T K M binos rectos conſtituit *per 22 tertii.*

Fig. IV.

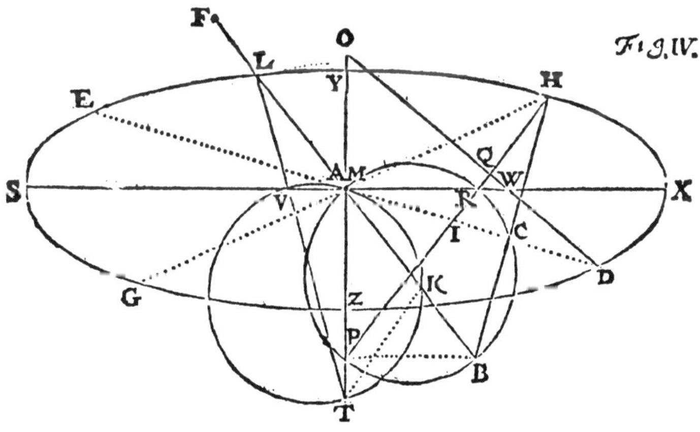

[6] *per 20 tertii*, atque in caſu fig. III per eandem K T V, propterea quòd uterque [6] angulo B A R ſeu K A V ſive
F A V æqualis eſt [f]; ſintque porrò, ob æquales angulos P A B,
T A K [7], tam peripheriæ P B, T K, quàm earum ſubtenſæ, nem-
pe

& 32 *tertii.* [f] In caſibus fig. IV & V tam angulus B P R ſeu B A R, quàm angulus K T V cum angulo K A V duos rectos conſtituit *per 13 primi, & 22 tertii.* [7] *per 26 & 29 tertii.*

[217] namely the sides PB and TK, opposite to the said equal angles, are mutually equal.[g] Thus it is clear that as the straight lines BC and PR, being produced, intersect at H, so the straight lines KM and TV, being produced, will intersect as well and in fact at the same point L, since TL equals PH. Indeed, from the preceding it follows[1] that the triangles BPH and KTL are similar and totally equal, and so the side KL equals the side BH. However,[2] the chord KM is also equal to the chord BC[h] since the angles KAM and BAC are equal. Therefore LM will be equal to HC too.

As the generator is KM, namely equal to BC, and as it has been applied in the angle KAM (which is the same as BAC,[i] or its opposite[k], or finally adjacent to it[l]) or eventually coincides with one of its arms and as, according to what has been proved, the intervals HB and HC are are also equal to LK and LM, therefore it follows that the point L –arbitrarily chosen on the given ellipse– and so the whole ellipse $SYXZ$, lies on the ellipse that has been described in the angle BAC with intervals HB and HC and so they coincide with each other at all points.

[g]in the case of Fig. III, in which the straight line BAF touches the circle TKV, the sides BP and TK are equal because the angles BAP and TVK are equal by III, 32. In the case of Fig. VI, in which the straight line PAY touches the circle TKV, the chords BP and TK are equal because the angles PAB and TMK are equal by III, 32;[1]I, 26;[2] III, 26 and 29;[h] in the case of Fig. III, KM is equal to BC, because the inscribed angle on the arc KTM is equal to the angle FAM or BAC by III, 32. In the case of Fig. IV, KM is equal to BC, because then the angle KTM is equal to the angle KAC or BAC by III, 32. In the case of Fig. V, KM is equal to BC because then the inscribed angle on the arc BC would be equal to the angle KAM, because then this angle and the other angle would constitute two right angles with the angle CAB by I, 13 and III, 22;[i] as in the case of Fig.VI;[k] as in the cases of Figs. I and II;[l] in the case of Fig. V;[m] as in the cases of Figs. III and IV.

pe latera P B, T K dictis æqualibus angulis adjacentia inter se æqualia [g] : apparet sicuti rectæ B C, P R productæ concurrunt in H, ita quoque rectas K M, T V productas, & quidem, cum ipsi P H æqualis sit T L, in ipso puncto L concursuras. quippe ex antedictis [1] similia atque in totum æqualia sunt triangula B P H, K T L, adeoque & latus K L lateri B H æquale. Est autem [2] & subtensa K M subtensæ B C æqualis [h], ob æquales an-

[g] In casu fig. III ubi recta B A F tangit circulum T K V, æqualia sunt latera B P, T K ob angulos B A P, T V K æquales *per 32 tertii.* In casu fig. VI, ubi recta P A Y contingit circulum T K V, æquales sunt subtensæ B P, T K ob angulos P A B, T M K æquales *per 32 tertii.* [1] *per 26 primi.* [2] *per 26 & 29 tertii.* [h] In casu fig. III, K M ipsi B C est æqualis, quandoquidem angulus qui consisteret in segmento K T M æqualis foret angulo F A M seu B A C *per 32 tertii.* In casu fig. IV, K M ipsi B C est æqualis, quandoquidem angulus K T M æqualis est angulo K A C seu B A C *per 32 tertii.* In casu fig V, K M ipsi B C est æqualis, quandoquidem angulus in segmento B C æqualis foret angulo K A M, utpote cum tam hic quàm ille cum angulo C A B duos rectos constitueret *per 13 primi & 22 tertii.*

Fig.V.

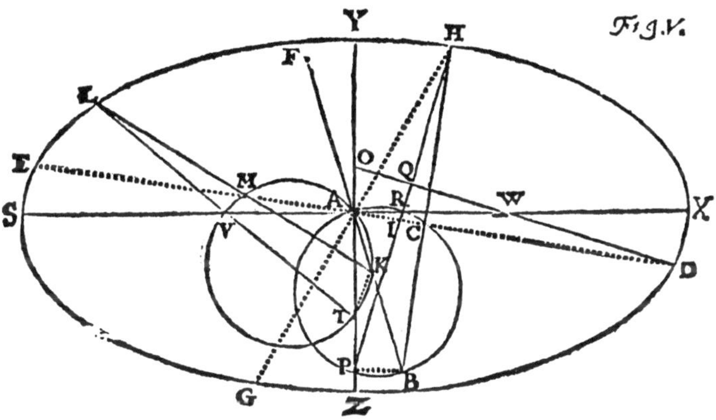

gulos K A M, B A C. Quocirca & L M ipsi H C æqualis erit. Unde cum describens sit K M, utpote ipsi B C æqualis, ac constituta in angulo K A M (qui cum ipso B A C vel idem [i], vel ei ad verticem [k], vel denique ipsi deinceps est [l]) aut certè cum alterutro crurum coïncidens [m], atque ex demonstratis æqualia quoque sint intervalla H B, H C intervallis L K, L M: sequitur punctum L, in exposita Ellipsi utcunque sumptum, id est, totam Ellipsin S Y X Z, esse in Ellipsi, quæ in angulo B A C, intervallis H B, H C, describitur, ideoque alteram alteri per

[i] ut in casu fig. VI.
[k] ut in casibus fig. I. & II.
[l] in casu fig. V.
[m] ut in casibus fig. III. & IV.

E c omnia

[218] However,[1] DE and HG are its (i.e., the second ellipse, transl.) conjugate diameters. Therefore they will also be conjugate diameters of the ellipse that is the same (i.e.,the given ellipse, transl.) and, as we know, DE is the transverse diameter and HG the second. This was to be proved.

Corollary 1

From this we not only conclude that all ellipses have their own axes, but also infer how to find, when arbitrary conjugate diameters are given, the axes of the ellipse to which these diameters belong [3.16].

E.g.: let DAE and HAG be conjugate diameters of an arbitrary ellipse. Then we first draw HB, equal to the semidiameter DA or AE and perpendicular to DE; next we connect B and A, and we bisect this segment at N. If we then describe a circle with center N and radius NA or NB, intersecting the straight line passing through H and N at P and R, then the segments HP and HR will have the length of the semiaxes. If we measure these on both sides of the center A towards or through the points R and P with the same length, such as the whole SX and YZ, then they will represent in magnitude and in position the required axes of the same ellipse, whose conjugate diameters are DAE and HAG.

Indeed, we first draw PB and choose AO equal to AR, and so[2] also equal to PB, and then we draw DO, intersecting SX at W. Because, accordingly, the described circle also passes through C,[3] since the angle ACB is a right angle, therefore the angles PBH and OAD are equal, as each of them

[1]Prop. 13, p. [205] and its Cor. 7; [2]I, 4; [3]III, 31.

omnia congruere. Sunt autem [1] hujus conjugatæ diametri D E, H G. Quare & illius, quæ cum ipsa eadem est, conjugatæ dia-

[1] per 13 hujus, ejusque Cor. 7.

Fig. VI.

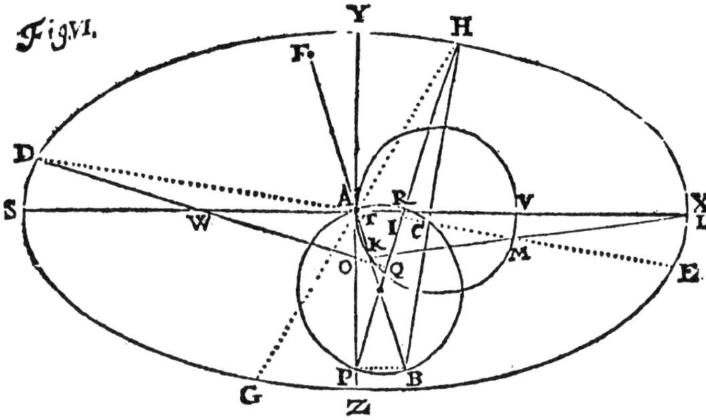

metri erunt, nimirum D E transversa, & H G secunda. Quod demonstrandum erat.

Corollarium. I.

Hinc colligitur non solùm Ellipses omnes suos habere axes, sed & quo pacto datis quibuslibet diametris conjugatis, ejus Ellipseos cujus diametri sunt, axes inveniantur.

Ut si cujuscunque Ellipseos conjugatæ diametri sint D A E & H A G, ductâ H B, semidiametro D A vel A E æquali, atque ad D E perpendiculari, junctâque B A ac ipsâ bifariam in N divisâ, si centro N intervallo N A vel N B circulus describatur, secans rectam per H & N ductam in P & R: erunt rectæ H P, H R semi-axes magnitudine, quæ idcirco utrinque à centro A versùs, aut per puncta R & P, æquali longitudine in directum positæ, sicut totæ S X & Y Z, exhibebunt magnitudine ac positione quæsitos axes ejusdem Ellipseos, cujus D A E & H A G conjugatæ diametri existunt.

Ductâ enim P B, sumptâque A O ipsi A R, ideoque [2] & ductæ P B æquali, agatur D O, occurrens ipsi S X in W. Cum itaque ob angulum A C B rectum descriptus circulus etiam per C transeat [3], erunt anguli P B H & O A D æquales, quoniam uterque

[2] per 4 primi.

[3] per 31 tertii.

[219] forms, together with the angle[a] PAC or PBC, two right angles.[1]

Since in the triangles OAD and PBH the sides OA and AD are equal to the sides PB and BH, respectively, and also lie around equal angles, therefore the base OD also[2] equals the base PH, i.e., the segment SA or AX, so that the angle AOD also equals the angle BPH or[3] PRA. From this follows that the side AW also[5] equals the side AP, and the side OW equals PR, because the triangles RAP and OAW are equal, since the angles at R and O are equal, and[4] RAP and OAW are right angles, and the sides RA and OA are equal.

Since[6] OW and PR are generators – being placed in reverse position– (p. [214]) of the ellipse whose axes are SX and YZ, with active points D and H, therefore it is clear from the foregoing demonstration, that the ellipse, described with axes SX and YZ, is in all respects the same as the ellipse whose conjugate diameters are DE and HG.

In this way it is clear that what has been proposed and proved by means of the foregoing theorem about an ellipse, described on arbitrary axes, is also applicable to an arbitrary ellipse, described on arbitrary conjugate diameters.

---

[a]viz., with angle PAC in the case of Fig. I and the like and with the angle PAC or PBC in the case of Fig. II and the like; [1]I, 13 and III, 22; [2]I, 4; [3]I, 29; [4]III, 31 and I, 13; [5]I, 26; [6]Prop. 13, p. [205].

que cum angulo [a] P A C seu P B C duos rectos constituit [1]. Unde cum triangula O A D, P B H latera O A, A D lateribus P B, B H, utrumque utrique, & quidem circa æquales angulos æqualia habeant: erit quoque [2] basis O D basi P H, id est, rectæ S A vel A X, ut & angulus A O D angulo B P H seu [3] P R A æqualis. Hinc cum æqualia sint triangula R A P, O A W, propter angulos ad R & O æquales, atque [4] R A P, O A W rectos, nec non latera R A & O A æqualia: erit etiam [5] latus A W lateri A P, ut & latus O W ipsi P R æquale. Quocirca cum [6] *describentes* sint

[a] nimirum *cum angulo* P A C in casu fig. I. & similibus, & *cum angulo* P A C *seu* PBC, in casu fig. II. & similibus.

[1] *per* 13 *primi &* 22 *tertii.* [2] *per* 4 *primi.* [3] *per* 29 *primi.* [4] *per* 31 *tertii &* 13 *primi.* [5] *per* 26 *primi.* [6] *per* 13 *hujus.*

Fig. I.

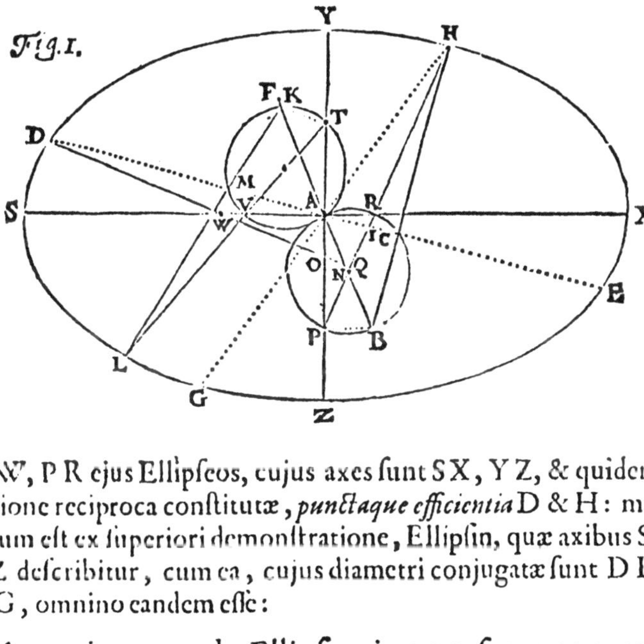

O W, P R ejus Ellipseos, cujus axes sunt S X, Y Z, & quidem in statione reciproca constitutæ, *punctaque efficientia* D & H: manifestum est ex superiori demonstratione, Ellipsin, quæ axibus S X, Y Z describitur, cum ea, cujus diametri conjugatæ sunt D E & H G, omnino eandem esse:

Atque ita, quæ de Ellipsi, circa quoscunque axes descripta, superiori Theoremate proposita ac demonstrata sunt, etiam cuilibet Ellipsi, & circa quascunque diametros conjugatas descriptæ, convenire, manifestum est.

[220]

Corollary 1

Furthermore, it follows from the proof of the same theorem, that in an ellipse all diameters are bisected by the center. Indeed, it has been proved that the part AE of a diameter, arbitrarily drawn, is equal to the part DA, as both are equal to the interval HB [3.17].

Corollary 3

It is, moreover, clear that in an ellipse the transverse diameter of arbitrary conjugate diameters is also a second diameter, and conversely. E.g., if DE is the transverse diameter of the (pair of) conjugate diameters DE and HG, and HG the second diameter, then HG will also be a transverse diameter, as in an ellipse an arbitrary diameter[1] is a transverse diameter and possesses a second diameter, conjugate to it. The proof, however, that DE is also the second diameter, conjugate to HG, will be immediately clear if one compares the Figures I and II and only transposes the letters and makes the necessary changes.

Corollary 4

Therefore, a straight line, drawn through the endpoint of a transverse diameter, parallel to the second diameter or to the ordinate-wise applied segments, will touch the ellipse at this point and at no other point, and will lie totally outside the ellipse[2] [3.18].

---

[1]Prop. 14, p. [213] and its Cor. 1; [2]Prop. 13, Cor. 2, p. [209] and Prop. 14, Cor. 3, p. [220].

Corollarium 2.

Sequitur porrò ex demonſtratione ejuſdem Theorematis, in Ellipſi diametros omnes à centro bifariam ſecari. demonſtratum enim eſt, in diametro D E, utcunque ductâ, partem A E parti D A æqualem eſſe, cum utraque *intervallo* H B æqualis ſit.

Corollarium 3.

Patet inſuper in Ellipſi, quarumcunque diametrorum conjugatarum tranſverſam etiam ſecundam eſſe, & contra. Ut, ſi conjugatarum diametrorum D E, H G tranſverſa ſit D E, & H G ſe-

Fig. 11.

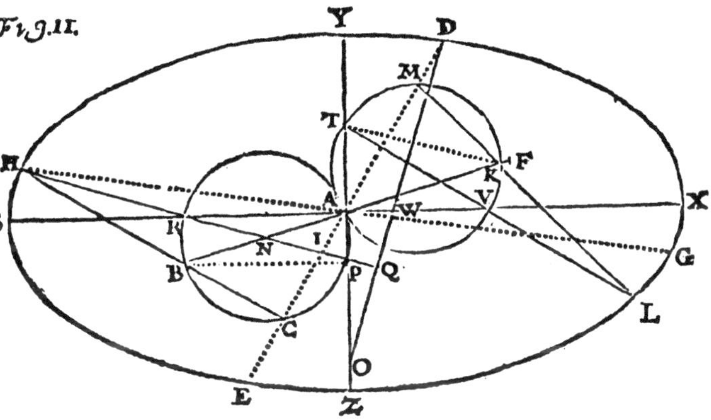

[1] per 14 hujus ejuſque Corol. 1.

cunda; cum in Ellipſi ducta quælibet diameter [1] tranſverſa ſit, habeatque ſecundam ſibi conjugatam, erit quoque H G tranſverſa. At verò & D E ſecundam eſſe ipſi H G conjugatam, factâ collatione figuræ I cum I I tranſpoſitis tantùm literis, ac mutatis mutandis demonſtratum ſimul apparebit.

Corollarium 4.

[2] per 2 Cor. 13. & 3 Cor. 14 hujus.

Quare & quæ per terminum tranſverſæ diametri ſecundæ æquidiſtans ſeu ordinatim applicatis parallela ducitur: Ellipſin in eodem termino & in nullo præterea puncto contingit, totaque extra Ellipſin cadit [2].

Co-

[221]

Corollary 5

And so an arbitrary straight line segment, from an arbitrary point on the curve ordinate-wise applied to an arbitrary diameter of an ellipse, will lie totally inside the ellipse, as it cannot lie totally outside the ellipse[1] nor intersect it at more than two points[2] [3.19].

Theorem XIV

Proposition 15

A line segment, connecting two arbitrary points on an ellipse, that is bisected by a diameter, will be drawn through the center or it will be ordinate-wise applied to this diameter, i.e., parallel to its conjugate diameter.

Indeed, in the ellipse $ABCD$, with center K, let the line segment EHG be bisected by the diameter AKC, where this segment does not pass through the center and is not parallel to the conjugate diameter BD. Then the following would be true: if we first apply GIF ordinate-wise (to AKC, transl.) and then draw the straight line GKL through the center, then the straight line through F and E and the straight line drawn through E and L would be one straight line, parallel to AC,[5] for GI is to IF[3] and GK is to KL[4] as GH is to HE. So it (FEL, transl.) would be ordinate-wise applied[6] to the other conjugate diameter, namely BD, and it would intersect the ellipse at three points. It has been proved above[7] that this is not possible.

Corollary 1

Therefore the following is true: if a diameter bisects an arbitrary line segment that has been drawn in an ellipse, but not through the center, then it will also bisect all segments parallel to this segment[8] [3.20].

[1] by the preceding Cor.; [2] Prop. 13, Cor. 5, p. [211]; [3] Prop. 13, Cor. 4, p. [211]; [4] Prop. 14, Cor. 2, p. [220]; [5] VI, 2; [6] Prop. 13, p. [205] and Prop. 14, Cor. 3, p. [220]; [7] Prop. 13, Cor. 5, p. [211]; [8] Prop. 13, Cor. 4, p. [211] and Prop. 15, p. [221].

Corollarium 5.

Adeoque quælibet recta, à quovis curvæ puncto ad quamcunque Ellipseos diametrum ordinatim applicata, tota intra Ellipsin cadit; utpote cum ea nec in totum extra Ellipsin cadere [1], nec eidem in pluribus quàm duobus punctis occurrere [2] possit.

[1] *per Cor. præcedens.*
[2] *per 5. Cor. 13. hujus.*

THEOREMA XIV.

Propositio 15.

Quæ bina quælibet Ellipseos puncta conjungens recta linea bifariam à diametro dividitur, erit aut per centrum ducta, aut ad eandem diametrum ordinatim applicata, hoc est, conjugatæ diametro æquidistans.

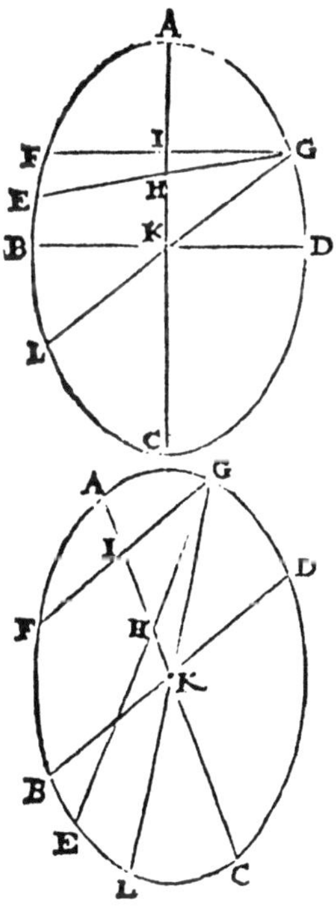

Si enim in Ellipsi ABCD, cujus centrum K, à diametro AKC bifariam divideretur recta EHG, quæ neque per centrum transeat, neque conjugatæ diametro BD æquidistans sit; applicatâ ordinatim GIF, ductâque per centrum rectâ GKL: Quoniam esset, ut GH ad HE, ita tam GI ad IF [3], quàm GK ad KL [4], recta per F & E, nec non per E & L ducta foret una linea recta diametroque AC parallela [5]; ideoque ad alteram ipsi conjugatam, nempe ad BD, ordinatim applicata [6], atque Ellipsi in tribus punctis occurreret; quod fieri non posse supra [7] ostensum est.

[3] *per 4 Cor. 13 hujus.*
[4] *per 2 Cor. 14 hujus.*
[5] *per 2 sexti.*
[6] *per 13 & 3 Cor. 14 hujus.*
[7] *in 5to Coroll. 13 hujus.*

Corollarium 1.

Ideoque si diameter rectam quamlibet in Ellipsi non per centrum ductam bifariam dividat, omnes quoque ipsi æquidistantes bifariam secabit [8].

[8] *per 4 Cor. 13. & 15tam hujus.*

[222]

Corollary 2

For this reason the following holds: if in an ellipse two arbitrary line segments have been drawn, mutually parallel, then the straight line that bisects both will pass through the center of this ellipse, i.e., it will prove to be one of its diameters. In fact, a diameter that is drawn through the midpoint of one of the parallels will also pass through the midpoint of the other parallel.[1] From this it is clear how to find as many diameters of a given ellipse as one likes, finding at the same time the straight lines that are ordinate-wise applied to them, as well as its center, for this is the common point of intersection of two or more diameters, and so it is also possible to find conjugate diameters and the axes[2] [3.21].

Corollary 3

From the foregoing it is easy to understand that an arbitrary line segment, connecting two arbitrary points on an ellipse, lies totally inside the ellipse.[3] Indeed, either it is[4] a diameter or it is ordinate-wise applied to the diameter that is drawn through its midpoint and the center [3.22].

Problem II

Proposition 16

To find in an arbitrary ellipse a diameter, conjugate to an arbitrarily drawn diameter.

Suppose that we have to find in a given ellipse $SYXZ$ another diameter, conjugate to an arbitrary diameter DAE. First we determine[5] the axes SAX and YAZ [3.23]; then from the endpoint D or E to one of the axes, e.g., to YAZ, we apply a line segment, e.g., DO, equal to the other semiaxis SA, which intersects, produced if necessary, this "other" axis, e.g., at W.

---

[1]Prop. 15, Cor. 1, p. [221]; [2]Prop. 14, Cor. 1, p. [218], or otherwise stated: as is obvious to everybody. [3]Prop. 14, Cor. 5, p. [221]; [4]Prop. 15 and its Cor. 1, p. [218]; [5]Prop. 15, Cor. 2, p. [222].

Corollarium 2.

Quocirca si in Ellipsi binæ quælibet rectæ sibi invicem æquidistantes ductæ sint, quæ utramque bifariam dividet recta linea per illius centrum transibit, seu ejusdem diameter existet. Quippe quæ per medium unius æquidistantium diameter ducetur per medium quoque alterius æquidistantium transibit [1]. Unde apparet, quo pacto datæ Ellipseos diametros quotlibet, simulque ad easdem ordinatim applicatas, nec non & ejus centrum, utpote quod duarum pluriumve diametrorum communis intersectio est, ideoque & diametros conjugatas, axesque [2] invenire liceat.

1 per 1 Cor. 15 hujus.

2 per 1 Cor. 14 hujus, aliterve, ut cuilibet obvium est.

Corollarium 3.

Ex dictis facilè apparet, quamlibet rectam, quæ bina quæcunque Ellipseos puncta conjungit, totam intra Ellipsin cadere [3]: utpote cum ipsa [4] vel diameter sit, vel ordinatim applicata ad eam diametrum, quæ per ipsius medium & centrum ducitur.

3 per 5 Cor. 14 hujus.

4 per 15 hujus ejusque Cor. 1.

PROBLEMA II.

Propositio 16.

In data quacunque Ellipsi ductæ cuilibet diametro alteram conjugatam invenire.

In data Ellipsi SYXZ ductæ utcunque diametro DAE altera conjugata invenienda sit.

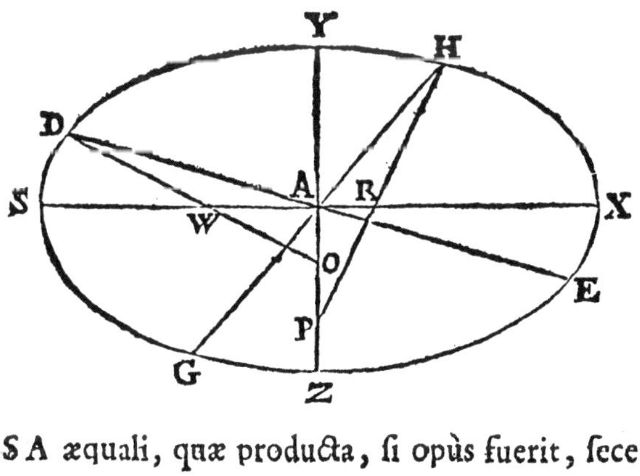

Inventis [5] axibus SAX & YAZ, atque à termino D vel E ad axium alterutrum, veluti ad YAZ, applicatâ rectâ, ut DO, semi-axi alteri SA æquali, quæ producta, si opùs fuerit, secet eundem axem alte-

5 per 2 Cor. 15 hujus.

[223] Next we apply the segment PR, in reverse position with respect to OW [3.24] and equal to it, so that of course AP and AR are equal to AW and AO, respectively and so that PR, being produced, intersects the ellipse at the point H, the endpoint, at the ellipse, of the segment HAG, drawn through the center A. It is clear–on the basis of what has been proved in Proposition 14 of this book– that HAG is the diameter, conjugate to DE.

And so it is at the same time clear that to separate diameters belong also corresponding different conjugate diameters, and that to one and the same diameter only one conjugate diameter can be constructed.

Corollary

From this it is furthermore clear how to draw, through an arbitrary given point on an ellipse, a straight line that touches the curve at this point and, apart from this point, at no other point. Indeed, if we first draw a diameter through the given point and the center, and if we find another diameter, conjugate to it,[1] and if we draw through the same point a straight line parallel to the constructed conjugate diameter, then this straight line[2] will be the required tangent.

Theorem XV

Proposition 17

At one and the same point there is –apart from the straight line, parallel to the diameter that is conjugate to the diameter, drawn through this point and the center– no other line that touches the ellipse.

Suppose that the ellipse $CHFG$ is touched at C by the straight line DCE, parallel to the diameter GH that is conjugate to the diameter CF, drawn through the point C and the center: then I say that no other straight line touches this ellipse at C [3.25].

Indeed, if this is possible, suppose then that the straight line ICK touches this ellipse at the point C as well and let another diameter NO be drawn, conjugate to the diameter LM that is parallel to this ICK. Since this NO differs from CF,[1] mentioned before,

[1]Prop. 16, p. [222]; [2]Prop. 14, Cor. 4, p. [220].

alterum, uti in W, applicetur in ſtatione reciproca ipſi O W, eidem æqualis recta P R, nempe ut A P, A R ipſis A W, A O ſingulæ ſingulis æquales ſint, ac producta P R Ellipſi occurrat in puncto H, à quo ſi per centrum A ducatur recta H A G, Ellipſi terminata: conſtat, per ea, quæ ad Propoſitionem 14^{tam} hujus libri demonſtrata ſunt, eandem H A G eſſe diametrum ipſi D E conjugatam.

Atque ita ſimul apparet, ſingulis diametris ſuas quoque diſtinctas conjugatas diametros eſſe, eidemque diametro unam tantum conjugatam duci poſſe.

Corollarium.

Unde porrò perſpicuum fit, quo pacto per datum quodlibet in Ellipſi punctum recta ducatur, quæ curvam in eodem ac in nullo alio præterea puncto contingat. Si enim ductâ per datum punctum & centrum diametro, inventâque alterâ ipſi conjugatâ [1], per idem punctum recta ducatur inventæ diametro conjugatæ æquidiſtans: erit eadem recta [2] contingens quæſita.

[1] *per 16 hujus.*
[2] *per 4 Cor. 14 hujus.*

T H E O R E M A XV.

Propoſitio 17.

Ellipſin in uno eodemque puncto præter rectam, quæ parallela eſt diametro illi, quæ per punctum & centrum ducitur, conjugatæ, alia recta non contingit.

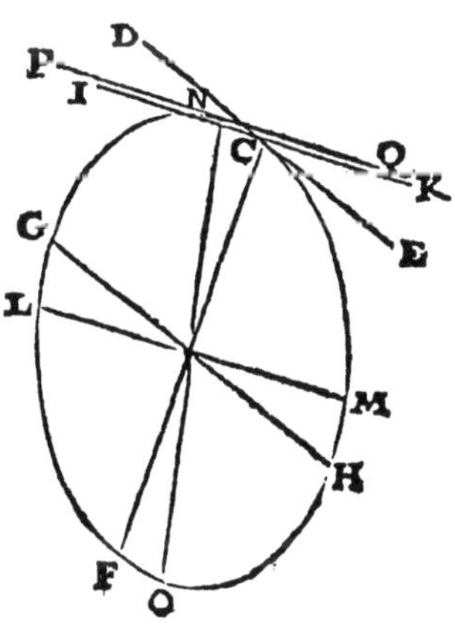

Contingat Ellipſin C H F G in puncto C recta D C E, parallela diametro G H, quæ conjugata ſit diametro C F, per punctum C & centrum ductæ: dico aliam rectam in puncto C eandem Ellipſin non contingere.

Si enim fieri poteſt, contingat eandem quoque in puncto C recta I C K, diametroque L M, eidem I C K æquidiſtanti, altera conjugata ducatur N O, (quæ cum à priori

C F

[224] the point N will not coincide with the point C. Let also be drawn PQ through N, parallel to LM and so also parallel to the tangent ICK. Then accordingly[2] the point C and so the straight line ICK will lie below the straight line PNQ, that is, towards the center of the ellipse. But in the same way[3] the point N and so the straight line PNQ lies also below the tangent ICK, i.e., towards the same center, which is a contradiction. So ICK does not touch the ellipse. This proof is valid for all other straight lines and so the proposition is true.

Corollary

Therefore it is clear[4] that in an ellipse the straight lines parallel to an arbitrary tangent are also parallel to the diameter conjugate to the diameter that is drawn through the point of contact and the center, that they are therefore ordinate-wise applied to the diameter, drawn through the point of contact, and that they are bisected by this diameter.[5] Conversely, (it is clear, transl.) the straight line drawn through the endpoint of an arbitrary diameter, parallel to an arbitrary straight line that is bisected by this diameter, will be tangent to the ellipse at this vertex.

Theorem XVI

Proposition 18

If an arbitrary tangent meets an arbitrary diameter of an ellipse (being produced) and if from the point of contact a line segment is applied ordinate-wise to this diameter, then the rectangle contained by the parts of the diameter, cut off, measured from the center, by the tangent and the applied line, will be equal to the square on the semidiameter, and conversely [3.26].

Suppose that the straight line DE touches an arbitrary ellipse GD, with center A, at the arbitrarily chosen point D and meets the diameter IG at E; from the point of contact D, let DC be applied ordinate-wise to this diameter: then I say that the rectangle CAE equals the square on the semidiameter AG.

Indeed, first let the diameter IG be an axis and then let OW be the generator in the position in which it was when it described the point D, so that the interval OD equals the semiaxis AG, and let PR be a generator compared with OW, in its reverse position (page [214]).

---

[1]Prop. 14, Cor. 4, p. [220], [2]Prop. 13, Cor. 2, p. [209]; [3]by the same Cor.; [4]by the preceding Cor.; [5]Prop. 13, Cor. 4, p. [211].

[1] per 4 Cor. 14 hujus. [2] per 2 Cor. 13 hujus. [3] per idem Coroll.

CF diversa sit, punctum N cum puncto C non coïncidet,) ac per N ipsi LM, ideoque & contingenti ICK, æquidistans ducta sit PQ. Cadet itaque [2] punctum C, adeoque recta ICK infra rectam PNQ: nimirum, versùs Ellipseos centrum. At verò & eodem modo [3] punctum N, ideoque recta PNQ, infra contingentem ICK: nempe, versùs idem centrum cadet, quod repugnat. Non contingit ergo ICK Ellipsin. Eadem de omnibus aliis est demonstratio, ac proinde constat propositum.

Corollarium.

[4] per Coroll. præcedens. [5] per 4 Cor. 13 hujus.

Constat itaque [4] in Ellipsi cuilibet tangenti parallelas, æquidistantes quoque esse diametro conjugatæ ei, quæ per tactum & centrum ducitur; ac proinde & ad diametrum per tactum ductam ordinatim applicari, atque ab illa bifariam dividi [5], & contra, quæ per cujuscunque diametri terminum ducitur æquidistans cuilibet rectæ, per eandem diametrum bifariam sectæ, Ellipsin in eodem vertice contingere.

THEOREMA XVI.

Propositio 18.

Si quælibet contingens productæ Ellipseos diametro cuicunque occurrat, atque à puncto contactus ad eandem diametrum recta ordinatim applicetur: erit rectangulum sub diametri portionibus, à centro per contingentem applicatamque abscissis, semidiametri quadrato æquale, & contra.

Quamcunque Ellipsin GD, cujus centrum A, contingat in puncto D, utcunque sumpto, recta DE, diametro IG occurrens in E; atque à puncto contactus D ad eandem diametrum ordinatim applicata sit DC: dico rectangulum CAE quadrato semidiametri AG æquale esse.

Sit enim primùm axis diameter IG, sitque OW *describens*, in statione uti fuit, cùm per eandem descriptum est punctum D; ita ut OD *intervallum* semi-axi AG æquale sit, PR autem *describens* in statione, ipsi OW reciprocâ; ita ut à curvæ puncto H, quod nempe

[225] Then the diameter HA, drawn from the point H on the curve, lying on the produced part of the generator PR, is conjugate to the diameter through D and A[1] and so also parallel to the tangent DE.[2] Next, let HF be ordinate-wise applied to the second axis AK and let OB and RT be drawn, parallel to AG and AK, intersecting the applied DC and HF–produced, if necessary–at B and T.

As the triangles OAW and RAP[3] are similar, therefore the triangles WCD and RTH will be similar too, as well as OBD and PFH.[4] But the sides WD and RH are also equal, as are OD and PH.[5]

Therefore the sides WC and RT, or AF, as well as DB and HF[6] will be equal too. Furthermore[7] the triangles EDC and HAF are equiangular. Therefore, because of what has been said before,[8] DC is to CW or to AF, i.e.,[9] EC is to HF or to DB, as the same DB is to BO.[10]

As EC, DB and BO form a continued proportion, therefore[11] the square on DB will be to the square on BO as EC is to BO or to CA and, by composition,[12] the square on DB[13] will be to the square on BO, i.e., the square on GA, will be to the square on CA, as EA is to CA; and so[14] the segments EA, GA, and CA form a continued proportion as well, and thus[15] the rectangle CAE equals the square on the semiaxis AG.

Since at the point D, no other straight line apart from DE, can touch the ellipse,[16] therefore it is clear that the converse is also true, namely: if the rectangle CAE equals the square on the semiaxis AG and if the straight line that has been ordinate-wise applied from C meets the ellipse at D, then the straight line connecting E and D is a tangent.

Next assume that the segment IG is not an axis of the ellipse GD, but another arbitrary diameter whose parameter is IB, and, from

---

[1]Prop. 14, Cor. 4, p. [220]; [2]Prop. 17, Cor.; p. [224]; [3]because of the construction; [4]I, 29 and VI, 21; [5]because of the construction; [6]I, 26; [7]I, 29; [8]VI, 4; [9]because the triangles EDC and HFA have the same angles; [10]because the triangles DCW and DBO have the same angles; [11]VI, 20, Cor.; [12]V, 18; [13]I, 47; [14] VI, 20, Cor.; [15]VI, 17; [16]Prop. 17, p. [223].

nempe *describenti* P R in directum est, ducta diameter H A conjugata sit ei, quæ per D & A duceretur [1], ideoque & contingenti D E parallela [2]. Sitque porrò ad secundum axem A K applicata H F, ducanturque O B, R T ipsis A G, A K æquidistantes, quæ applicatis D C, H F, productis, si opus fuerit, occurrant in B & T.

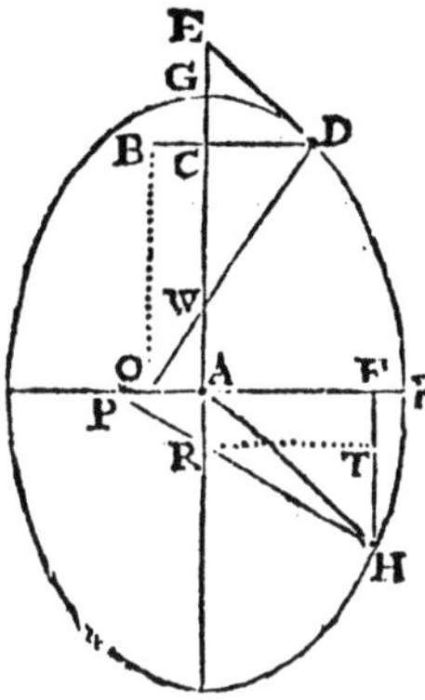

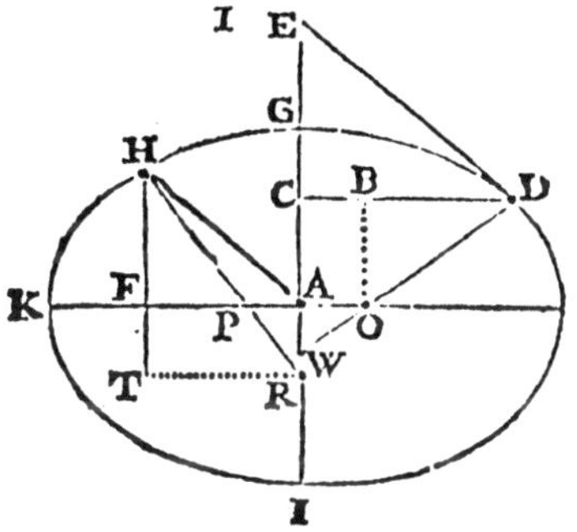

Itaque cum similia sint triangula O A W & R A P [3], erunt quoque triangula W C D & R T H, nec non O B D & P F H [4] similia. At verò & latera W D & R H, nec non O D & P H [5] æqualia sunt. Quare & latera W C & R T [6] sive A F, nec non D B, & H F [6] æqualia erunt. Sunt autem porrò [7] triangula E D C & H A F [8] æquiangula; unde ex antedictis erit [8] D C ad C W sive A F, id est [9], E C ad H F sive D B, uti eadem D B ad B O [10]. Unde cum proportionales sint E C, D B, B O, erit [11] ut E C ad B O sive C A, ita D B quadratum ad B O quadratum; & componendo [12], ut E A ad C A, ita [13] D O quadratum ad B O quadratum, hoc est, G A quadratum ad C A quadratum; ac proinde [14] & rectæ E A, G A, C A proportionales erunt, ideoque [15] rectangulum C A E quadrato semi-axis A G æquale. Cumque in puncto D alia recta præter ipsam D E Ellipsin contingere non possit [16], patet conversum quoque verum esse: nimirum, si rectangulum C A E æquale sit quadrato semi-axis A G, & per C ordinatim applicata Ellipsi ocurrat in D, junctam E D esse contingentem.

Deinde non sit recta I G Ellipseos G D axis, sed alia diameter quæcunque, cujus parameter I B, atque ab assumpto in curva

[1] per 4 Cor. 14 hujus.
[2] per Cor. 17 hujus.
[3] ex constructione.
[4] per 29 primi, & 21 sexti.
[5] ex constructione.
[6] per 26 primi.
[7] per 29 primi.
[8] per 4 sexti.
[9] propter triangula E D C, H F A æquiang.
[10] propter triangula D C W, D B O æquiangula.
[11] per Cor. 20 sexti.
[12] per 18 quinti.
[13] per 47 primi.
[14] per Cor. 20 sexti.
[15] per 17 sexti.
[16] per 17 hujus.

[226] an arbitrarily chosen point D on the curve, let DC be ordinate-wise applied to the same diameter, and suppose that the square on the semidiameter AG equals the rectangle CAE. Then I say that the line connecting E and D, being produced, lies totally outside the ellipse and that it therefore touches the ellipse at the point D, and conversely.

Indeed, on this ED or on its produced part, let an arbitrary point F be chosen ad libitum, and through F let there be drawn the straight line FH, parallel to CD, which meets the said diameter IG at H, but the ellipse GD at K (surely, if FH did not meet the ellipse, it is more than clear that the point F would lie outside the ellipse) [3.27].

Next we suppose that another ellipse –GL–has been described with IG as an axis and with the same parameter IB and that, through C and H, CL and HM are ordinate-wise applied to this same axis, intersecting the curve at L and M; we also suppose that E and L are connected by a straight line (which at any rate touches the ellipse GL at L[1]) and that this straight line, being produced if necessary, meets the produced HM at N. Since the square on DC is to the rectangle GCI as the square on LC is to the same rectangle GCI (in fact both ratios are the same as the ratio of the parameter IB to the diameter or to the axis IG[2]), therefore[3] the squares on DC and LC and so also the segments DC and LC will be equal.

In the same way it can be proved that the segments KH and MH are equal too.

But since CD is to HF as CL is to HN (since[4] both ratios are the same as the ratio of the segment EC to the segment EH), therefore also[5] HF and HN will be equal. However, HN is greater than the applied HM, as ELN is a tangent: therefore HF will also be greater than the applied HK and so the point F, arbitrarily chosen on the straight line EDF, i.e., the whole EDF will lie outside the ellipse GD or, equivalently, EDF will touch this ellipse at the point D.

Since it is not possible that –apart from EDF– another straight line touches the ellipse at the point D,[6] therefore the converse is also clear, namely: if ED touches the ellipse GD at D

[1]by what has been proved above; [2]Prop. 13, p. [205] and VI, 20, Cor.; [3]V, 9; [4]VI, 4; [5]V, 14; [6]Prop. 17, p. [223].

utcunque puncto D ad eandem diametrum ordinatim applicetur D C, sitque quadrato semidiametri A G æquale rectangulum C A E: dico junctam E D, productamque, totam extra Ellipsin cadere, ideoque eandem in puncto D contingere, & conversim.

Sit enim in eadem E D, aut in ipsa producta, prout libuerit, assumptum utcunque punctum F, sitque per F ducta recta F H ipsi C D æquidistans, quæ dictæ diametro I G occurrat in H, Ellipsi verò G D in K. (Etenim si Ellipsi non occurreret, manifestissimè punctum F extra Ellipsin foret.) Deinde I G ut axe, eâdemque parametro I B, intelligatur descripta alia Ellipsis G L, ac per C & H ad eundem axem ordinatim applicentur C L, H M, quæ curvæ occurrant in L & M, junganturque E L, (quæ utique Ellipsin G L in L continget [1],) eaque producta, si opùs fuerit, productæ H M occurrat in N.

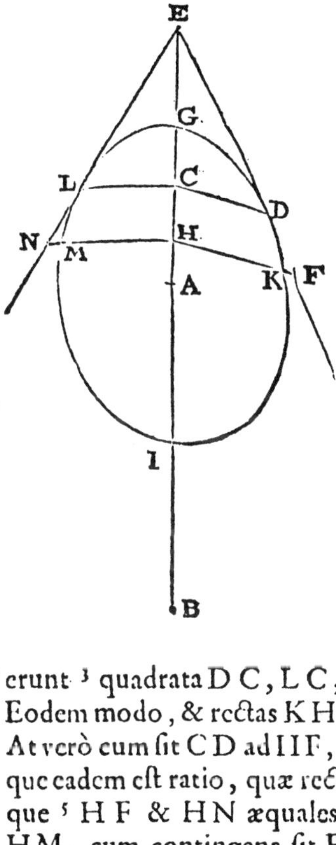

[1] per sup. demonstr.

Itaque quoniam est quadratum D C ad rectangulum G C I, ut quadratum L C ad idem G C I rectangulum, (quippe utriusque eadem est ratio, quæ parametri I B ad diametrum sive axem I G [2]:) erunt [3] quadrata D C, L C, ideoque & rectæ D C, L C æquales. Eodem modo, & rectas K H, M H æquales esse, demonstrabitur. At verò cum sit C D ad H F, ut C L ad H N, (siquidem [4] utriusque eadem est ratio, quæ rectæ E C ad rectam E H,) erunt quoque [5] H F & H N æquales. Est autem H N major applicatâ H M, cum contingens sit E L N: ergo & H F applicatâ H K major erit, ideoque punctum F, in recta E D F utcunque sumptum, hoc est, tota E D F, extra Ellipsin, G D cadet, sive, quod idem est, eandem in puncto D continget. Cumque non possit præter E D F alia recta eandem Ellipsin in puncto D contingere [6], manifestum quoque est conversum: si nempe E D Ellipsin G D in-

[2] per 13 hujus, & Cor. 20. sexti.
[3] per 9 quinti.
[4] per 4 sexti.
[5] per 14 quinti.
[6] per 17 hujus.

[227] and meets the diameter GI at E and if DC has been ordinate-wise applied to the same diameter, then the rectangle ACE equals the square on the semidiameter AG.

Corollary

From what has been said it is clear how to draw, from an arbitrarily given point, a straight line that touches an ellipse.

Indeed, if the given point lies on the curve itself, e.g., D, then we have already shown above[1] [3.28] how to draw the tangent at the said point. This can, however, also be done in the following way, by means of the foregoing theorem.

First we find a diameter[2] GI and from D we draw a line segment DC, ordinate-wise aplied to it; next we construct the rectangle CAE, equal to the square on the semidiameter AG and we connect E and D.

But if the given point, e.g., E, lies outside the ellipse, then we proceed as follows: we draw the straight line EA to the center A,[3] which will intersect the ellipse at G; we construct the rectangle EAC, equal to the square on AG; and through C we apply CD ordinate-wise, where CD[4] is of course parallel to the tangent at G[5] and intersects the ellipse at D. Then we connect E and D. This line will be exactly the required tangent, in the former case as well as in the latter case.[6]

It is very clear, however, that it is not possible to draw from a point inside an ellipse a straight line that touches this ellipse.

So I have –as I trust– transmitted briefly, sufficiently clearly, and in the most natural way, without considering any solid, the foundations and the most important properties of the curves that the Ancients called conic sections [3.29].

From these fundamental principles all other theorems pertaining to a parabola, a hyperbola, or an ellipse can be very easily deduced without further guidance by one who uses his brains sufficiently, and he can proceed by himself to a higher level in geometry.

I think it superfluous to dwell longer on this way of treating these topics, especially as there remains a prominent and more sublime science to which the Ancients applied themselves strenuously, as is clear from the reports of some of them and from several fragments of ancient geometers.

---

[1]Prop. 16, Cor., p. [223], [2]Prop. 15, Cor. 2, p. [222]; [3]found by Prop. 15, Cor. 2, p. [222]; [4]according to Prop. 17, Cor., p. [224]; [5]by the preceding text or by Prop. 14, Cor. 5, p. [221]; [6]Prop. 18, p. [224].

in D. contingat, diametroque G I occurrat in E, & ad eandem diametrum ordinatim applicata sit D C, rectangulum C A E quadrato semidiametri A G æquale esse.

Corollarium.

Ex dictis perspicuum est, quo pacto à dato quolibet puncto ducenda sit recta, quæ Ellipsin contingat.

Si enim datum punctum in ipsa curva sit, veluti D: jam supra [1] ostensum est, quo pacto per dictum punctum contingens ducatur. Quod tamen & hoc quoque modo per præcedens Theorema perficietur.

Ductâ ex D ad inventam [2] diametrum G I rectâ ordinatim D C, fiat rectangulum C A E quadrato semidiametri A G æquale, jungaturque E D.

At si extra Ellipsin sit datum punctum, ut E: ductâ ad A centrum [3] rectâ E A, quæ Ellipsin secet in G, quadrato A G æquale fiat rectangulum E A C; ac per C ductâ ordinatim applicatâ C D: nimirum, quæ [4] æquidistet contingenti quæ per G duceretur [5], occuratque Ellipsi in D, jungatur E D: eritque hæc ipsa tam priori quàm posteriori casu [6] contingens quæsita.

A puncto autem intra Ellipsin dato non posse duci rectam, quæ eandem contingat, manifestissimum est.

[1] in Cor. 16. hujus.
[2] per 2 Cor. 15. hujus.
[3] inventum per 2 Coroll. 15 hujus.
[4] juxta Cor. 17. hujus.
[5] per præcedentia, aut 5 Coroll. 14. hujus.
[6] per 18 hujus.

Atque ita me compendiosè viâ satis planâ ac maximè naturali, absque ulla solidi consideratione, Elementa proprietatesque præcipuas Curvarum, quas Veteres *Coni sectiones* appellavêre, tradidisse confido. E quibus principiis cætera omnia, quæ ad Parabolam, Hyperbolam, vel Ellipsin pertinent, absque ulteriori manuductione facillimè deducet, quicunque animum iis debitè applicuerit, atque in Geometricis per se ad ulteriora progredi valeat. Adeò ut eâdem tractandi methodo hisce diutiùs inhærere supervacuum putem, præsertim cum insignis & sublimior quædam scientia supersit, cui Veteres enixissimè incubuisse ex quorundam relatu ac nonnullis antiquorum Geometrarum fragmentis manifestum est; quæque tam ab iisdem

[228] This science was called by them, as well as by later geometers, "The Finding of Loci" or "Construction of Loci" [3.30].

It is certainly reliable that –in order to promote this science– Apollonius and the other geometers, especially those things put together in writing that they added to the said Foundations in their treatment of the conic sections.

I think that a deeper understanding of the curved lines, a complete enumeration thereof, and a classification, as well as a distribution into their genera and species [3.31] with a separation of those curves that in fact do not belong to Geometry from those that are to be included in geometry, can especially be obtained from a careful treatise on Loci [3.32]. Therefore I thought it useful to add this treatise here; however, not in the way initiated, it seems, by the Ancients. Indeed a complete and huge book would hardly be sufficient to this end, even if it should contain only an exact theory of the Loci that they called Plane and Solid (in my opinion, however, less correct) [3.33], i.e., the lines that appear as a straight line, as a Parabola, as a Hyperbola, as an Ellipse, or as the circumference of a circle (in our opinion they aimed only at the construction of these Loci). Furthermore this volume would develop into an immense volume if it would be extended to Loci [3.34] that are curved lines of the second genus, as is our intention.

I will, however, follow the analytic method [3.35] by means of the investigation of equations and general rules, by which all possible cases can be solved and determined.

In investigating them we will observe the order in which, after the explication of the foundations of the parabola, the hyperbola and the ellipse, we will transmit the invention and the determination of those loci that are either straight lines or that consist of the curves mentioned before.

Here we suppose knowledge of what pertains to the nature of straight lines, angles and rectilinear figures and also of circles. These, however, will in our writings too

quàm à Recentioribus *Locorum Inventio* sive *Compositio* appellata fuit. Ad quam promovendam, ab Apollonio cæterisque Geometris ea præcipuè conscripta esse, quæ in Conicorum tractatione prædictis Elementis superaddidere, omnino credibile est. Cumque penitiorem curvarum linearum notitiam perfectamque earum enumerationem ac distinctionem, ut & distributionem in sua genera & species, cum segregatione earum, quæ verè Geometricæ non sunt, ab iis quæ in Geometriam sunt recipiendæ, ex accurata *Loci* tractatione imprimis petendam existimem: è re fore duxi, eandem tractationem hîc subjungere, non quidem eâ methodo, sicut à Veteribus inchoata videtur, cum vix integrum & ingens volumen eidem sufficeret, si vel tantùm *Locorum*, quæ *Plana*, ac *Solida* (quamvis, meo judicio, minùs rectè,) vocârunt, id est, quæ vel *recta linea*, vel *Parabola*, vel *Hyperbola*, vel *Ellipsis*, sive *circuli circumferentia* existunt, (quorumque Locorum Compositioni eos solummodo intentos fuisse invenimus,) doctrinam exactè complecteretur, atque id porrò volumen in immensum excresceret, si ad Loca, quæ sunt lineæ curvæ secundi generis, uti nobis propositum est, extenderetur; sed Arte Analyticâ per Æquationum examen & præcepta generalia, quibus omnes omnino casus possibiles resolvantur ac determinentur: In quibus pertractandis cum ordinem simus observaturi, ut jam post explicationem Elementorum Parabolæ, Hyperbolæ, & Ellipsis, (suppositâ notitiâ eorum, quæ ad linearum rectarum, angulorum, & figurarum rectilinearum, nec non Circulorum naturam pertinent) inventionem ac determinationem tradamus eorum *locorum*, quæ vel rectæ lineæ sunt vel ex prædictis curvis constant; (Illa autem & nobis, ne quid temerè mutemus,

[229] go by the name of Plane Loci and Solid Loci to avoid a reckless change, and thereby we will show that to the curves of the first genus, distinct from the circle, only the parabola, the hyperbola, and the ellipse are to be admitted.

We will, however, preface the treatise on the higher loci, which belong to the curved lines of the second genus, in the same way with the foundations of these curves. The way to their generation, however, is not only paved by the description of the curved lines of the first genus, which have been introduced and explained in this book, but also by many other manners of describing them in the plane. Therefore we thought it worthwhile to add here at least those ways (which are certainly infinite in number, as everyone who is intent on these speculations easily will experience) that either will be helpful to us in describing curves of the second genus or –in our opinion– are more suited than the preceding ways, for the mechanical delineation in the plane of curves of the first genus [3.35].

Chapter IV

Another method to delineate in the plane a parabola, a hyperbola and an ellipse

Let ABC be an arbitrary isosceles triangle and let the equal sides AB and AC as well as the base BC be indefinitely produced on both sides, e.g., to D and E and to F and G as well as to H and I. Let also from one of the angles at the base be drawn an arbitrary, bounded line segment, parallel to the opposite side, e.g., BK, and let through its endpoint K another straight line, let us say $LAKM$ –indefinitely extended on both sides– be freely movable, rotating about the vertex of the remaining angle, namely the point A, as a pole. Finally, let CN, having one point on FG and parallel to DE, pass through the point of intersection C of FG and HI.

Then I say the following: if the angle EBH and its opposite angle DBI move together with the straight line BK in both directions, in such a way, however, that the arm AB always remains on the straight line DE and so that at the same time the straight line HI moves the straight line CN back and forth, always parallel to itself,

mus, *Locorum Planorum*, *Solidorumque* nomine venient) atque eo ipso ostendamus in primo curvarum genere, præter Circulum, non nisi Parabolam, Hyperbolam, & Ellipsin esse recipiendas. Tractationi autem ulteriorum locorum, quæ pertinent ad lineas curvas secundi generis, similiter quoque earundem curvarum Elementa præmittemus. Cum verò ad ipsarum generationem viam sternant non tantùm descriptiones linearum curvarum primi generis, hoc libro propositæ atque explicatæ, sed & multi alii illas in plano describendi modi: operæ pretium duximus eorundem modorum, qui certè infiniti sunt, ut quilibet huic speculationi intentus facilè experietur, vel illos saltem hîc adjungere, quos aut ad descriptiones curvarum secundi generis auxilio nobis fore, aut Mechanicæ curvarum primi generis in plano delineationi præcedentibus aptiores judicamus.

CAPUT IV.

Alia Parabolam, Hyperbolam, & Ellipsin in plano delineandi Methodus.

Sit triangulum quodcunque isosceles ABC, & tam æqualia crura AB, AC, quàm basis BC utrinque indefinitè producantur, ut ad D, E, & F, G, nec non HI; sitque ab alterutro angulorum ad basin ducta quævis recta terminata, opposito cruri æquidistans, ut BK, & per terminum ejusdem K altera recta, utrinque indefinitè extensa, liberè transeat, quæ circa verticem anguli reliqui, nempe punctum A, ut Polum, circulariter mobilis sit, veluti LAKM; ac denique rectæ FG insistens CN ipsi DE parallela transeat per ipsarum FG & HI intersectionem C. Dico, si angulus EBH atque ipsi ad verticem DBI cum recta BK moveatur in utramque partem, ita tamen ut crus AB semper applicatum maneat rectæ DE, simulque recta HI huc atque illuc promoveat rectam CN, sibi ipsi semper æquidistantem,

[230] while the straight line BK rotates the said LM (which always passes through the point K) about the pole A –then the point of intersection of CN and LM, which is O, will describe a parabola with diameter AD and parameter KB, while FG touches this parabola at the vertex A [4.1].

Indeed, let the angle EBH or DBI be in an arbitrary position; then if the point of intersection of the straight lines FG and HI is denoted by C and if from the point of intersection O, the segment OP is applied to the diameter [4.2], parallel to FG, then KB will always be to BA, i.e., to AC, as the same AC is to CO[1] and so[2] the rectangle contained by KB and CO, i.e.,[3] contained by KB and AP, will be equal to the square on AC, i.e.,[4] on OP. Therefore: if BAC is a right angle, then AD will be the axis; if not, it is a diameter and the ordinate-wise applied segments will form angles with it, equal to the angle BAC or BAG.

In passing it must be remarked here also that, by the same motion, the point of intersection of HI and LM, say Q, describes a hyperbola or two opposite hyperbolas.

[1]I, 29 and VI, 4; [2]VI, 17; [3]I, 34; [4]as [3]

tem, ac recta B K ad polum A circulariter moveri faciat prædictam L M, per punctum K ſemper tranſeuntem, interſectionem ipſarum C N, L M, quæ ſit ad O, Parabolam deſcribere,[1] cujus diameter eſt A D, parameter K B, ac F G eandem contingens in vertice A.

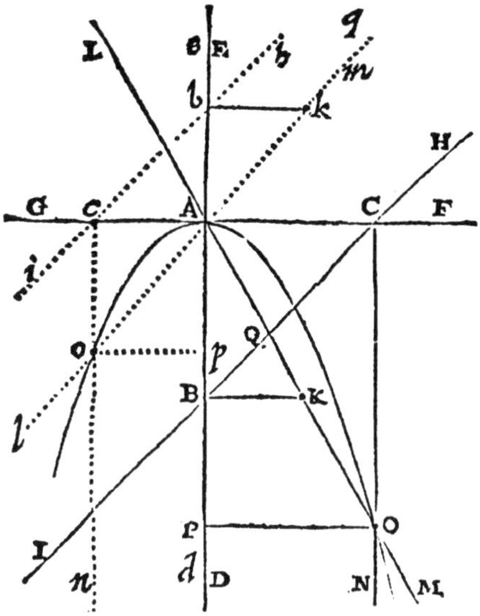

In quacunque enim ſtatione conſtitutus fuerit angulus E B H ſeu D B I, ſi interſectio rectarum F G, H I deſignetur per C, atque ab interſectionis puncto O ad diametrum applicata ſit O P ipſi F G æquidiſtans: erit ſemper K B ad B A, hoc eſt, ad A C, uti eadem A C ad C O[1]: ac proinde [2] rectangulum ſub K B, C O, id eſt [3], ſub K B, A P quadrato rectæ A C, hoc eſt [4], ipſius O P æquale. Unde ſi B A C angulus rectus fuerit, erit A D axis, ſin minùs, diameter, ad quam ordinatim applicatæ faciunt angulos ipſi B A C vel B A G angulo æquales.

[1] per 29 primi, & 4 ſexti.
[2] per 17 ſexti.
[3] per 34 primi.
[4] per eandem.

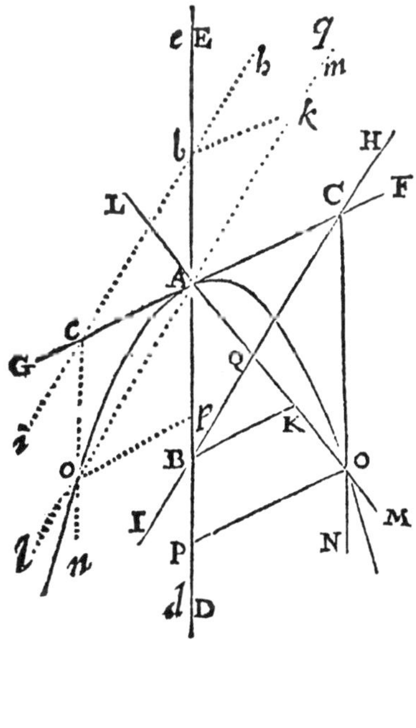

In tranſitu etiam hîc notandum eſt, eodem illo motu per interſectionem ipſarum H I, L M, puta Q, Hyperbolam ſive oppoſitas

[231] It must also be remarked that, even if the triangle BAC is not isosceles or if the segment BK is erected not from the vertex B but from an arbitrary point on AD, the curve AO would nevertheless be a parabola. In the first case, however, the parameter would not remain the same, nor the vertex and the diameter in the second case; but in these cases too they would be very easy to determine.

At the end of the first chapter we pointed out that the curve, described according to the definitions given at the beginning of the first chapter, with an arbitrary active line and an arbitrary interval, is a hyperbola if the moving angles are not equal to the angles at the directrix at the same side (of the interval, transl.) and we think this rather useful for a mechanical delineation of this curve in the plane [4.3].

Therefore we thought that we must establish this now by a demonstration, while we will at the same time show how this method can be easily applied to the foresaid delineations of hyperbolas.

Let then the curve DAM be described with the active line IG, with the interval AL, and with the directrix KLO, but with unequal angles IAL and KLA. Then I say that this curve is a hyperbola (Figure 1, page [232], transl.); I also say the following: if we first draw the line segment AK from the pole A to the directrix, so that the angle LAK equals the angle LAG, and if we describe a circle with center A and radius AK, intersecting the active line at I and G and the directrix at K and Q,[a] and if we draw through the points I and K as well as through G and Q the straight lines IK and GK, intersecting each other at F, then the straight lines FI and FG are the asymptotes[b] [4.4].

Indeed, let us first take an arbitrary point on the curve, e.g., D, and let us bring the moving angle, e.g., OAD, as well as the generator, e.g., OD, into the position in which they were when they described the point D.

Since the angles AIK and AKI are now mutually equal[1] and also, if taken together, equal to the angle KAG (in fact, the latter as well as the first two together, form together with the angle IAK two right angles), therefore the angles AIK– or AIF– and GAL will also be mutually equal, as they are halves of equal angles, and for this reason[3] the straight lines IKF and AL are parallel,[c] and therefore, as IKF meets the straight line FG, so the generators AL and OD

---

[a]or touching it at K, as has been shown in the case of Fig. V; [b]if, however, two of the said points coincide, like I and K in Fig. III and G and Q in Fig. IV, then the straight line touches the circle there, as one sees IF touching in the first case, and GF touching in the second case; [1]I, 5; [2]I, 13 and 32; [3]I, 28; [c]in the case of Fig. III, the straight lines IF and AB will be parallel, because each angle AIF and GAL is a right angle.

positas Hyperbolas describi; ut &, quamvis triangulum B A C isosceles non foret, nec etiam recta B K ex angulari puncto B sed ubivis in recta A D educta esset, nihilominus tamen curvam A O Parabolam fore; at verò nec parametrum priori, nec verticem, nec diametrum posteriori casu easdem remanere, quas tamen illis quoque casibus determinare facillimum est.

Quoniam autem circa finem capitis primi monuimus, curvam , juxta definitiones in principio ejusdem capitis propositas, quâlibet *efficiente*; & quocunque *intervallo* descriptam, si *anguli mobiles* inæquales sint *iis qui ad directricem* sunt ab eadem parte, Hyperbolam esse, idque Mechanicæ ejusdem in plano delineationi non inutile judicamus: idcirco id demonstratione jam comprobandum duximus, simul ostensuri, quo pacto eadem Methodus ad prædictas Hyperbolarum delineationes commodè applicetur.

Sit itaque *efficiente* I G, *intervallo* A L, & *directrice* K L O, angulis autem I A L & K L A inæqualibus, descripta curva D A M: dico eandem curvam Hyperbolam esse; ac si ductâ à *Polo* A ad *directricem* rectâ A K, ita ut angulus L A K angulo L A G æqualis sit, centro A & intervallo A K circulus describatur, secans *efficientem* in I & G, ad *directricem* in K & Q [a], perque puncta I & K, nec non per G & Q ducantur rectæ I K, G Q, sibi mutuò occurrentes in F, rectas F I, F G Asymptotos esse [b].

Sumpto enim in curva puncto utcunque, veluti D, applicetur tam *angulus mobilis*, ut O A D, quàm *describens*, ut O D, in statione uti fuêre, cùm per eas descriptum est punctum D. Quoniam igitur æquales sunt anguli A I K, A K I inter se [1], nec non simul sumpti angulo K A G, (quippe tam posterior quàm priores [2] cum angulo I A K binos rectos constituunt): erunt quoque anguli A I K seu A I F & G A L, utpote æqualium dimidia, inter se æquales, ac propterea [3] rectæ I K F & A L parallelæ [c]; ideoque sicut I K F rectæ F G occurrit, ita & eidem F G occurrent

[a] aut eandem in K contingens, uti in casu fig. V. exhibito.

[b] si verò dictorum punctorum bina coïncidant velut I & K in III, ac G & Q in IV. fig. tangat ibidem circulum recta, ut I F in priori, & G F in posteriori cum contingere cernitur.

[1] *per 5 primi.* [2] *per 13 & 32 primi.* [3] *per 28 primi.* [c] in casu fig. III, quoniam uterque angulorum A I F & G A L rectus est, rectæ I F, A B parallelæ erunt.

[232] will meet the same FG too, as they are parallel to IKF. Now suppose they meet FG at B and C and suppose we draw through B the straight line BN, parallel to the directrix KO and meeting the generator DO at N: then as GA is equal to AI, so[1] GB will be equal to BF.

However, since in the triangles LAK and OQC the angles at L and O are equal (because[2] AL and CO are parallel) and also because the angle LAK or GAL, i.e., GIF, is equal to OQC (in fact, each of them forms, together with the angle KQG or KQE,

[1]VI, 2 and V, 14; [2]I, 29.

rent describentes AL & OD, utpote ipsi IKF æquidistantes. Sint itaque ipsarum occursus in B & C, ac per B agatur recta BN directrici KO æquidistans, occurrensque describenti DO in N: eritque ut GA ipsi AI, ita [1] GB ipsi BF æqualis. Cum

[1] *per 2 sexti, & 14 quinti.*

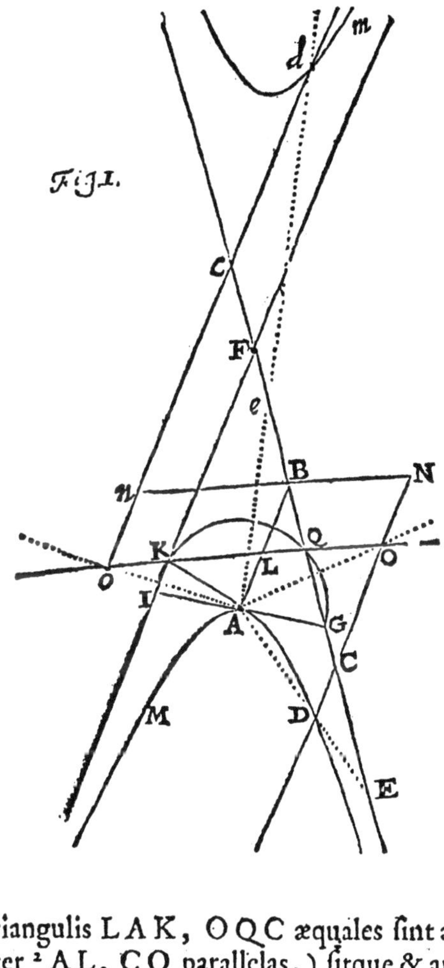

autem in triangulis LAK, OQC æquales sint anguli ad L & O, (propter [2] AL, CO parallelas,) sitque & angulus LAK sive GAL, id est, GIF, æqualis angulo OQC, (quippe tam hic quàm ille cum angulo KQG, vel KQE duos rectos constituit

[2] *per 29 primi.*

[233] two right angles[1]),[d] therefore these triangles LAK and OQC or[2] NBC will be equiangular. Moreover, the following holds: the angle AGE equals the angle IKO or ALO (in fact, each of them forms together with the angle IGQ or IGF[3] two right angles), and if one adds to the angles LAG and OAD (or OAE), which are identical or equal, the common part OAG[f] – or if one subtracts this–then the sums or the differences LAO and GAD (or GAE) are also equal and LO and AO meet each other. Therefore it is necessary that GE and AD meet each other too. Let E be their point of intersection; then the triangles AGE and ALO will be equiangular too, and therefore[4] AL will be to AG as LO, or NB, is to GE.

---

[1]in Figs. I and II because of I, 13 and III, 22; in Fig. IV because of I, 13 and III, 18 and 31; [d]in the case of Fig. III it appears that both angles OQC and LAK are right angles, because of I, 13 and III, 31 and in Figs. V and VI it appears that the angles GIF and OQC are equal, because of III, 32 and III, 21; [2]I, 29; [3]in Fig. I because of I, 13 and III, 22; in Fig. III because of I, 13 and III, 31; in Fig. IV because of I, 13 and III, 18 and 31; in Fig. V because of I, 13 and III, 32; in Fig. VI because of I, 13, because, by III, 21, IKQ equals the angle IGQ; [e]in Fig. II the angle AGE equals the angle ALO because each of them forms with the angle IKQ two right angles by I, 29 and III, 22; [f]or–in the case of Fig. II and similar –BAE; [4]because of VI, 4 in a different order.

tuit[1],) [d] æquiangula erunt eadem triangula LAK & OQC ſive [2] NBC. Porrò, quoniam angulus AGE angulo IKO ſeu ALO æqualis eſt, (quippè tam hîc quàm ille cum angulo IGQ ſive IGF [3] binos rectos conſtituit [e],) atque angulis LAG, OAD vel OAE iiſdem ſive æqualibus addito vel ablato communi OAG [f], compoſiti vel reſidui LAO, GAD vel GAE æquales quoque ſunt, ac LO, AO ſibi mutuò occurrant; GE quoque & AD ſibi mutuò occurrant neceſſe eſt; ſit itaque ipſarum occurſus E punctum; & æquiangula erunt triangula, AGE, ALO, eritque propterea [4] AL ad AG, ut LO ſive NB ad GE. At verò

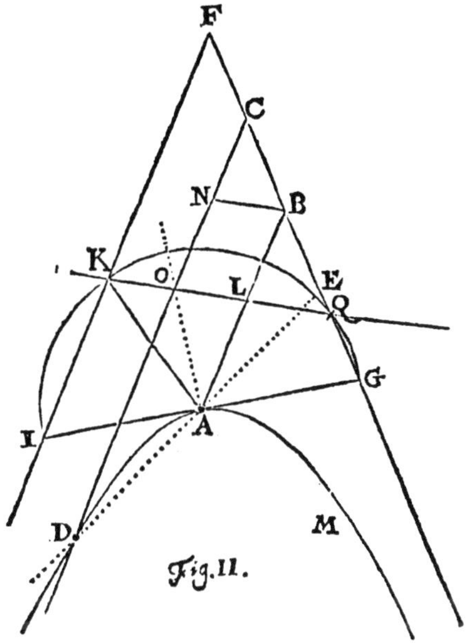

Fig. II.

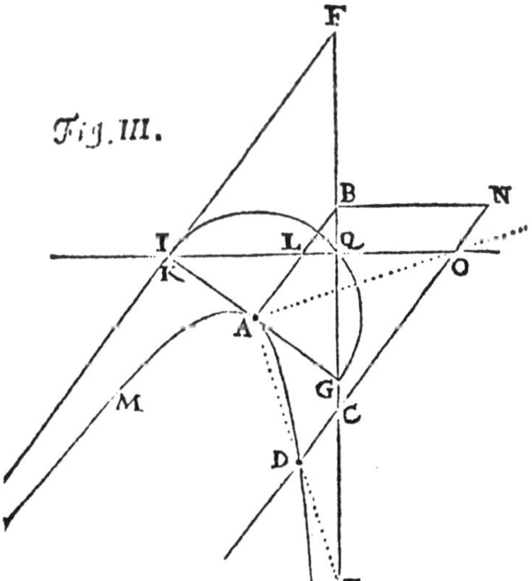

Fig. III.

[1] in I & II fig. per 13 primi, & 22 tertii. in IV per 13 primi, & 18 ac 31 tertii.

[d] in caſu fig. III apparet, tam angulum OQC quàm LAK rectum eſſe, per 13 primi, & 31 tertii : ac in fig. V & VI angulos GIF & OQC æquales, per 32 tertii & 21 ejuſdem.

[2] per 29 primi.

[3] in fig. I, per 13 primi & 22 tertii, in fig. III, per 13 primi & 31 tertii; in fig. IV per 13 primi, 18 & 31 tertii; in fig. V per 13 primi & 32 tertii; in fig. VI, per 13 primi, quoniam angulo IGQ æqualis eſt IKQ per 21 tertii.

[e] in fig. II angulus AGE angulo ALO eſt æqualis, quia uterque cum angulo IKQ duos rectos conſtituit, per 29 primi & 22 tertii.

[f] vel, in caſu fig. II & ſimilibus, BAE.

[4] per 4 ſexti permut.

[234] But (as the triangles LAK and NBC are similar[1]) the same AL is also[2] to AK, i.e., to the same AG, as NB is to BC. From this follows[3] that NB will be to GE as the same NB is to BC. Therefore[4] the segments GE and BC will be equal and so also GB (or[5] FB) and CE will be equal, as well as BE and FC. Finally: because the triangles ABE and DCE[6] are similar, BE[7] is to CE, i.e.,[8] FC is to FB, as BA is to CD.

Therefore[9] the rectangle FCD, contained by the extremes, will be equal to the rectangle FBA contained by the means. As this is always the case

---

[1]because of what has been proved above; [2]VI, 4; [3]V, 11; [4]V, 9; [5]because of what has been proved above; [6]I, 29; [7]VI, 4; [8]because of what has been proved above; [9]VI, 16.

1 per sup. dem. (ob triangula LAK & NBC [1] similia) est quoque [2] eadem AL ad AK, hoc est, ad eandem AG, ut NB ad BC. unde per consequens erit [3], ut NB ad GE, ita eadem NB ad BC. ac proinde [4] rectæ GE & BC, ideoque & GB seu [5] FB & CE, nec non BE & FC æquales erunt. Denique cum, propter triangula

2 per 4 sexti.
3 per 11 quinti.
4 per 9 Quinti.
5 per sup. demonstr.

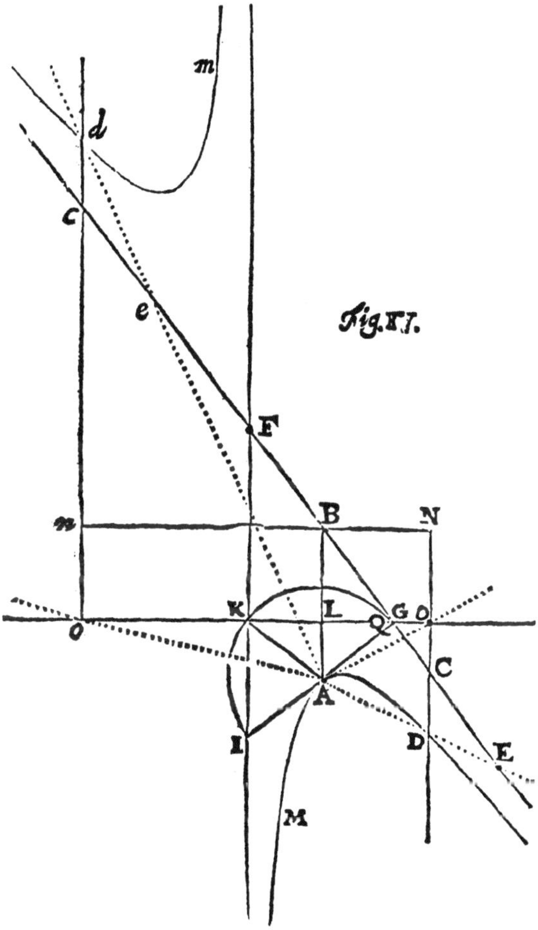

Fig. XI.

ABE & DCE [6] similia, [7] BE sit ad CE, hoc est [8], FC ad FB, ut BA ad CD: erit [9] rectangulum FCD sub extremis æquale rectangulo FBA sub mediis. Quod cum semper accidat, ubi-

6 per 29 primi.
7 per 4 sexti.
8 per sup. demonstr.
9 per 16 sexti.

[235] wherever on the curve the point D has been chosen, it follows[1] that the curve DAM is a hyperbola, with asymptotes FI and FG. Which was to be proved.

From what has been said, the following is clear: if the active line (in the initial position, transl.) or the tangent[2] (at the vertex, transl.), as IG, is perpendicular to one of the the asymptotes, as in the third and the fourth figure, then either the moving angles LAI and LAG will be right angles (in the case that the interval, e.g., AL, has been drawn parallel to the asymptote that is intersected by the active line or tangent IG at right angles, as in the third figure) or certainly the generator will be perpendicular to the directrix (in the case that the interval is parallel to the asymptote that is intersected by the active line or tangent GI at oblique angles, as in the fourth figure) [4.5].

Now suppose either that given the asymptotes FI and GG and the tangent IG,

[1]Prop. 3, p. [180]; [2]Prop. 6, p. [190].

ubicunque in curva assumptum fuerit D punctum, sequitur [1] curvam D A M Hyperbolam esse, cujus Asymptoti F I, F G. Quod erat ostendendum. [1] *per 3 hujus.*

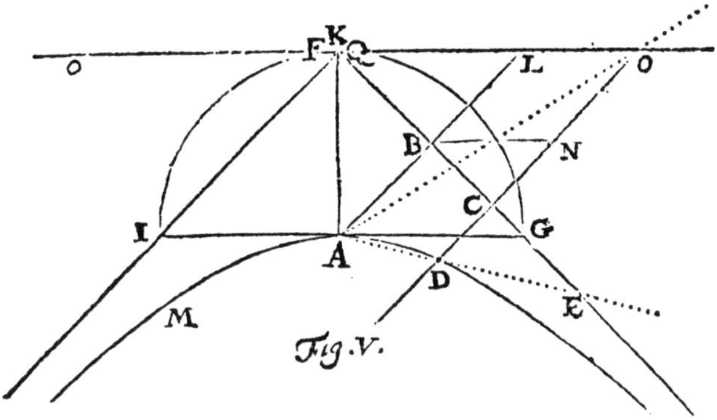

Fig. V.

Ex antedictis manifestum est, si *efficiens* seu [2] contingens, ut I G, ad Asymptotorum alterutram perpendicularis sit, veluti in tertia & quarta figura, vel *angulos mobiles* L A I & L A G rectos fore, si nempe *intervallum*, ut A L, æquidistans ductum sit ei Asymptoto cui *efficiens* seu contingens I G ad angulos rectos occur- [2] *per 6 hujus.*

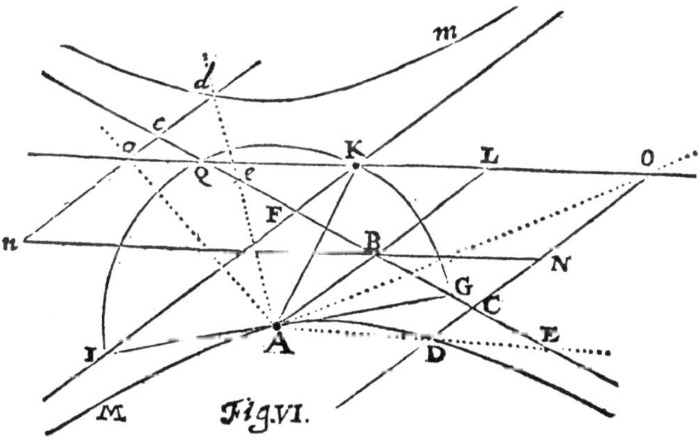

Fig. VI.

rit, ut in tertia figura, vel certè *describentem* ad *directricem* fore perpendicularem, si nempe *intervallum* parallelum fuerit ei Asympto, cui eadem *efficiens* seu contingens G I occurrit ad angulos obliquos, ut in quarta figura.

Itaque si vel Asymptotis F L, F G, & contingente I G; vel diametris

[236] or given the conjugate diameters HA and IG, a hyperbola is to be described [4.6].

In the latter case we first draw the asymptotes FI and FG and we describe a circle, with diameter IG, that intersects both asymptotes, say at K and Q.[a] Next we draw through K and Q the straight line KO, which is intersected by the straight line AL, drawn parallel to one of the asymptotes, e.g., FI, at L. It is very easily seen, on the basis of what has been said, that the curve described, with the active line IG, the interval AL and the directrix KO, will be the same as the hyperbola that was to delineate.

Sometimes, however, it will be profitable to achieve the same without describing the circumference of a circle, to avoid the intersection of a circumference and straight lines at oblique angles.

---

[a]or touches one straight line and intersects the other one, as happens in Fig. II at I and Q and in Fig. III at G and at K.

metris conjugatis H A, I G, Hyperbola ſit deſcribenda, ductis in caſu poſteriore Aſymptotis F I, F G, diametro I G circulus deſcribatur, qui ſecet utramque Aſymptoton, puta

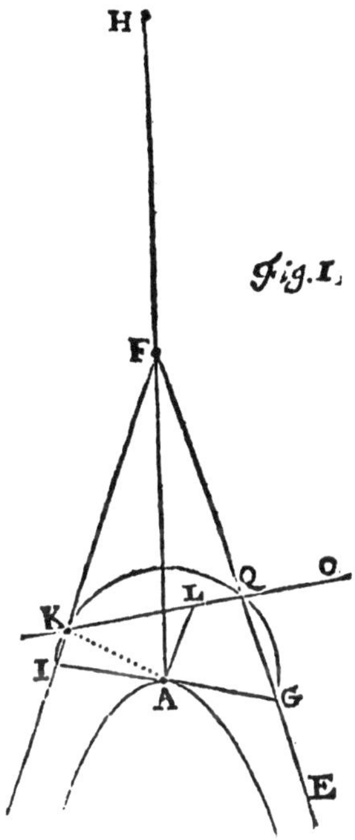

Fig. I.

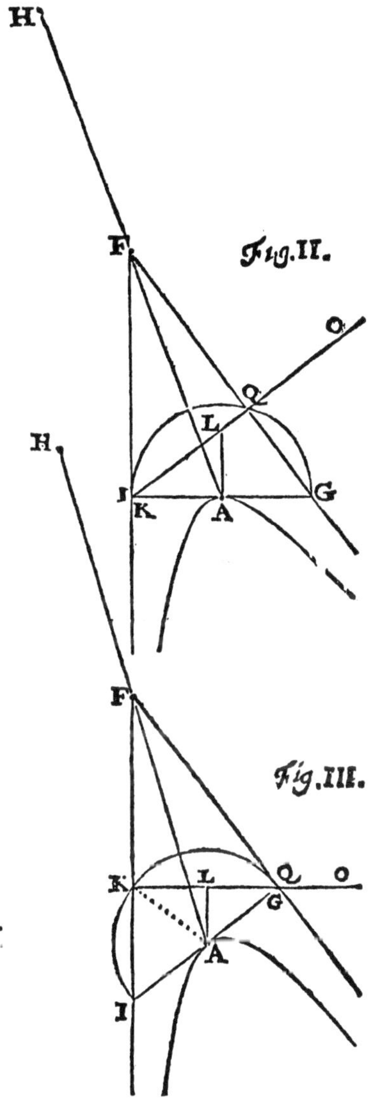

Fig. II.

Fig. III.

in K & Q ª, ductâque per K & Q rectâ K O, cui ducta A L, Aſymptotorum alterutri, ut F I, æquidiſtans, occurrat in L: facillimè colligitur ex præmiſſis, ſi *efficiente* I G, *intervallo* A L, ac *directrice* K O, curva deſcribatur, eandem fore Hyperbolam, quæ delineanda proponitur.

ª aut alteram tangat, & alteram ſecet, ut in II fig. fit in I & Q, ac in III fig. in G & K.

Nonnunquam tamen, ut obliquos circumferentiæ & rectarum occurſus evitemus, hæc eadem abſque Circuli deſcriptione efficere expediet. Ita-

[237] This goes as follows [4.7]: if we first draw AL parallel to one of the asymptotes, e. g., to FI, and if we then draw AK to the same asymptote in such a way that the angle LAK equals the angle LAG, and if we draw through K the straight line KO intersecting the said AL at L, so that the angle FKO equals the angle FGI, then the curve described, with the active line IG, the interval AL, and the directrix KO, will be exactly the required hyperbola.

For mechanical descriptions of hyperbolas we thought it furthermore rather useful to show here with a few words how arbitrary hyperbolas can be delineated in the plane if the moving angles are right angles or if the generator is perpendicular to the directrix [4.8].

Suppose that in the first or second case a hyperbola is to be described in the plane, with asymptotes FS and FT, touched by the segment ST, bounded by the asymptotes on both sides [4.8].

First draw from one of the points S or T a straight line perpendicular to one asymptote or the other, e.g., TV, which is supposed to form right angles with FT. Through the point I (which has been chosen in such a way that IF is the mean proportional between VF and SF) draw IG, parallel to the same TV, which will also touch the required hyperbola,[1] because VF is to IF, i.e., [2] IF is to SF, as[3] TF is to GF.

If one describes a circle IKG on this IG (as a diameter, transl.) that touches the asymptote FT at G[4] and intersects the other asymptote at K, and if one draws the straight line KGO through K and G, which is intersected at L by the line AL, drawn from A, the midpoint of the tangent IG (AL being perpendicular to this IG or to KO), then the hyperbola described, with the active line IG, the directrix KO, and the interval AL (perpendicular to this active line or to the said directrix), is –according to the prescriptions that have just now been exposed– exactly the hyperbola that has to be delineated.

Likewise it is not difficult to delineate an arbitrary hyperbola in the plane

[1]Prop. 9, p. [195]; [2]by hypothesis; [3]VI, 2 and after "composition," by V, 18; [4]III, 16, corollary.

Itaque si, ductâ A L Asymptotorum alterutri, ut F I, parallelâ, ad eandem Asymptoton ducatur A K, ita ut L A K angulus angulo L A G æqualis sit, & per K recta K O secans prædictam A L in L, ita ut angulus F K O angulo F G I æqualis sit: erit curva, *efficiente* I G, *intervallo* A L, ac *directrice* K O descripta, ea ipsa Hyperbola, quæ quæritur.

Ad Mechanicas porrò Hyperbolarum descriptiones non inutile fore judicavimus paucis hîc ostendere, quo pacto vel *angulis mobilibus* rectis, vel ita ut *describens* ad *directricem* sit perpendicularis, quælibet Hyperbolæ in plano delineari queant.

Si itaque vel hoc, vel illo modo describenda sit in plano Hyperbola, cujus Asymptoti sint F S, F T, quamque contingat recta S T, utrinque Asymptotis terminata: Ductâ ab alterutro punctorum S vel T rectâ, vel ad hanc, vel ad illam Asymptoton perpendiculari, uti T V, quam ad F T angulos rectos efficere supponimus, eidem T V per punctum I (nempe ita sumtum ut I F inter V F & S F media sit proportionalis) agatur æquidistans I G, quæ continget quoque Hyperbolam quæsitam [1], propterea quòd sit V F ad I F, hoc est [2], I F ad S F, uti [3] T F ad G F. Ideoque descripto super eandem I G circulo I K G, qui tangat [4] Asymptoton F T in G, atque alteram secet in K, si per K & G ducatur recta K G O, eidemque occurrat ducta ab A, puncto medio tangentis I G, recta A L in L, quæ quidem A L vel ad eandem I G, vel ad ductam K O sit perpendicularis: erit Hyperbola, quæ *efficiente* I G, *directrice* K O, atque *intervallo* A L, ad eandem *efficientem*, dictamvè *directricem* perpendiculari, describitur, juxta ea quæ modò exposita sunt, hæc ipsa, quæ delineanda proponitur.

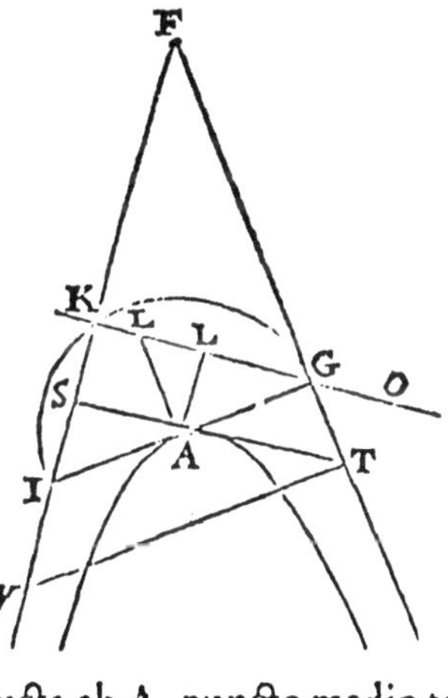

[1] *per 9 hujus.*
[2] *ex hypothesi.*
[3] *per 2 sexti, & componendo per 18 quinti.*
[4] *per Cor. 16 tertii.*

Similiter & vel datis quibuslibet *angulis mobilibus*, vel

[238] when arbitrary moving angles are given or when the generator forms arbitrary angles with the directrix.

For the rest it will sometimes be useful that the following way to describe an ellipse in the plane has also been added.

Let a line segment, e.g., ABC, rotating about the pole B by means of two of its points A and C (be it that B lies between A and C or that C lies between A and B), move forward the straight lines ADE and DCF, always parallel to themselves [4.9] (first figure on page [238], transl.) and intersecting each other at right angles.

Then I say that the the curve that is described by their continuously moving point of intersection, e.g., D, is an ellipse with center B and axes GBH and IBK, that is to say, with magnitude twice AB and BC, lying, however, on parallels to ADE and DCF, drawn through the pole B and bisected at this point [4.10].

Indeed, let us take on this curve an arbitrary point, e.g., D, and let us bring the generators ADE and DCF in the position in which they were when they described the point D by their intersection (second figure on page [238], transl.); let us moreover denote the point at which one of them, e.g., ADE, intersects one of the straight lines GH and IK or the other, e.g., GH, say at L, and let GAH be the circumference of the circle that is described by the motion of the point A. Since[1] the square on AB is now to the square on BC –i.e., the square on GB is to the square on BK– as the square on AL, or[2] the rectangle GLH, is to the

[1]VI, 2 and 22; [2]II, 14 or III, 35.

ita ut *deſcribens* ad *directricem* datos quoſlibet angulos efficiat quamcunque Hyperbolam in plano delineare haud difficile erit.

Cæterùm ſequentem quoque Ellipſin in plano deſcribendi rationem hîc adjeciſſe ſuum aliquando uſum habebit.

Recta linea, ut A B C, ad Polum B circulariter mota binis ſui punctis A & C, in eadem utcunque aſſumptis (ſive B ſit inter A & C, ſive C ſit inter A & B,) promoveat rectas A D E, D C F, ſibi ipſis ſemper æquidiſtantes, ac ſe invicem ad rectos angulos interſecantes: dico curvam, quæ continuâ earundem interſectione, veluti D, deſcribitur, Ellipſin eſſe, cujus centrum eſt B, & axes G B H, I B K, nempe magnitudine ipſarum A B, B C duplæ, poſitione verò ipſis A D E, D C F, æquidiſtantes per B Polum ductæ, atque ibidem bifariam diviſæ.

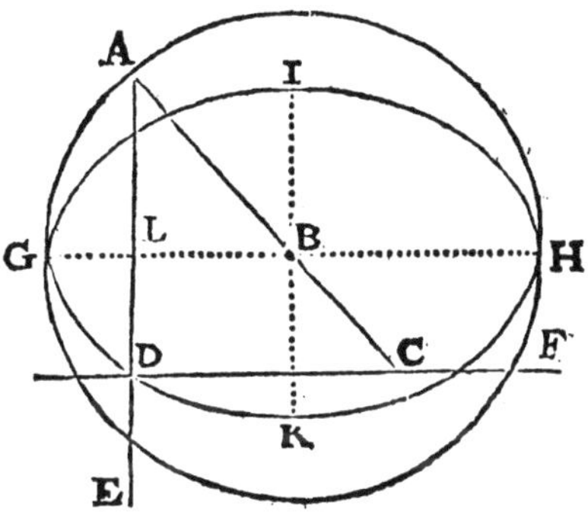

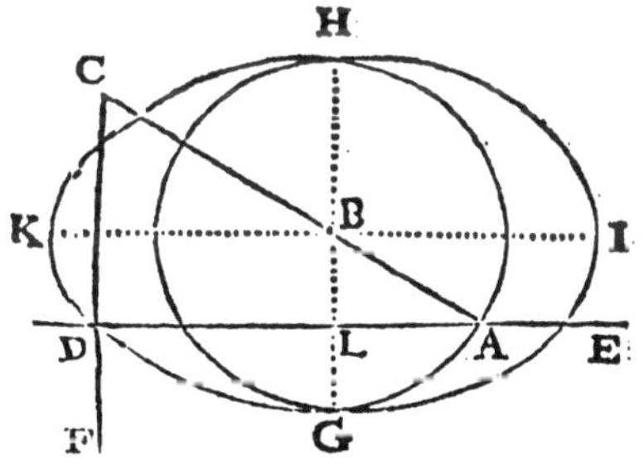

Sumpto enim in eadem curva puncto utcunque, veluti D, applicentur ipſi *deſcribentes* A D E, D C F in ſtatione uti fuêre, cùm per illarum interſectionem deſcriptum eſt punctum D; noteturque porrò punctum, ubi earum alterutra, veluti A D E, vel hanc vel illam ductarum G H, I K, ex. gr., ipſam G H, ſecat, ut in L. & ſit G A H circumferentia Circuli, qui per motum puncti A deſcribitur. Quoniam itaque eſt [1] A B quadratum ad B C quadratum, hoc eſt, G B quadratum ad B K quadratum, ut A L quadratum ſive [2] G L H rectangulum ad L D qua-

[1] *per 2 & 22 ſexti.*
[2] *per 14 ſecundi, vel 35 tertii.*

[239] square on LD, therefore it is clear[1] that the curve GKH, described in the way that has been mentioned before, is an ellipse whose axes are GH and IK.

It is, however, clear that, if the points A and C have the same distance to the pole, the said curve will be the circumference of a circle.

Next let ABC be not one straight line, but an arbitrary angle ABC, either obtuse or acute, and let the aforementioned straight lines DAE and DCF be drawn through the points A and C in such a way that, one of them

[1] Prop. 13, p. [205].

quadratum: conſtat [1], curvam G K H, uti prædictum eſt, deſcriptam Ellipſin eſſe, cujus axes ſunt G H, I K.

[1] per 13 hujus.

Manifeſtum autem eſt, ſi puncta A & C æqualiter à B Polo diſtent, prædictam curvam Circuli circumferentiam fore.

Non ſit deinde A B C una linea recta, ſed angulus quicunque, ſive obtuſus, ſive acutus A B C, ſintque prædictæ rectæ D A E,

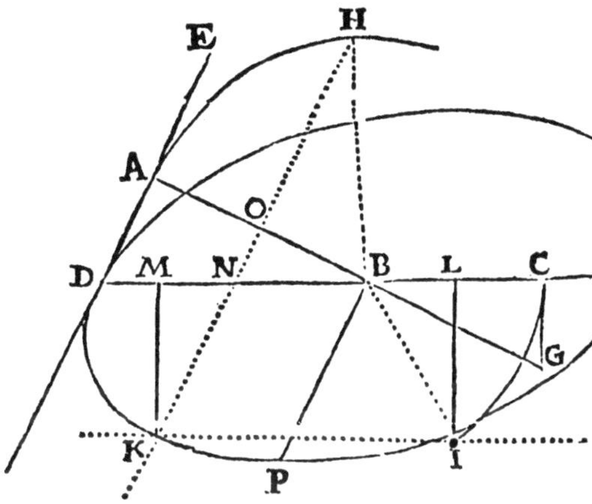

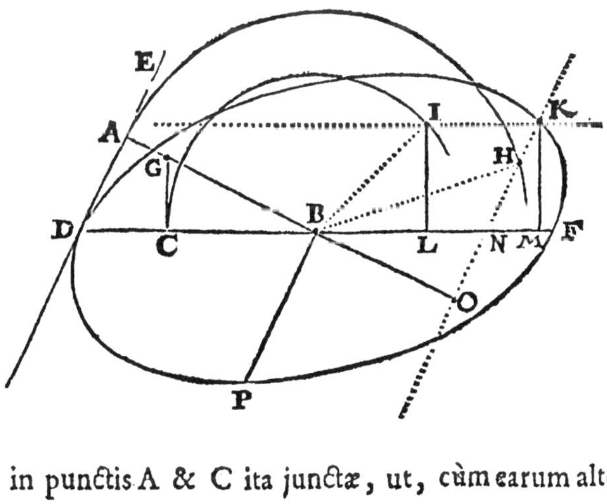

D C F in punctis A & C ita junctæ, ut, cùm earum altera uni

cruri

[240] coincides with one arm (as in the position of ABC the straight line DCF coincides with the arm BC), whereas the other is perpendicular to the remaining arm (as in the same position the straight line DAE is perpendicular to the arm AB).

Then I say again: if the angle ABC moves forward the straight lines DAE and DCF, always parallel to themselves, by means of the points A and C arbitrarily chosen on each of the arms and rotating about the pole B, then the curve described by their (scil. DAE and DCF, transl.) continuously moving point of intersection, e.g., D or K, is an ellipse, whose semidiameters have as lengths the segments DB and BG. These are namely the parts of the beforementioned arms (produced if necessary), intercepted between the perpendiculars AD and CG (respectively drawn through the chosen points A and C) and the pole [4.11]. One of these semidiameters, e.g., DB, even has the right position; however, the other, e.g., BG, has not, but it (scil. this semidiameter, transl.) is BP, equal to BG, but parallel to DAE [4.12].

Indeed, let the said angle ABC be in another arbitrary position, e.g., in the position of HBI; thus the aforementioned point of intersection will be at K. If, however, one drops from I and K the perpendiculars IL and KM on the segment DF (twice the segment DB) and if one denotes the points of intersection by N and O, then the following is true: as the angles ABC (or OBL) and HBI are equal or identical, therefore the angles HBO and IBL will be equal too, after adding or subtracting the common HBF; so the triangles HBO and IBL are equiangular, because the angles O and L are right angles.[1] However,[2]

---

[1]by hypothesis and I, 29; [2]VI, 21.

cruri coincidat, (quemadmodum in ſtatione A B C recta D C F coincidit cruri B C,) altera ad reliquum crus ſit perpendicularis, (ſicut in eadem ſtatione recta D A E ad crus A B perpendicularis eſt:) dico iterum, ſi angulus A B C circa Polum B circulariter motis punctis A & C in utroque crure utcunque aſſumptis promoveat rectas D A E & D C F ſibi ipſis ſemper æquidiſtantes, cur-

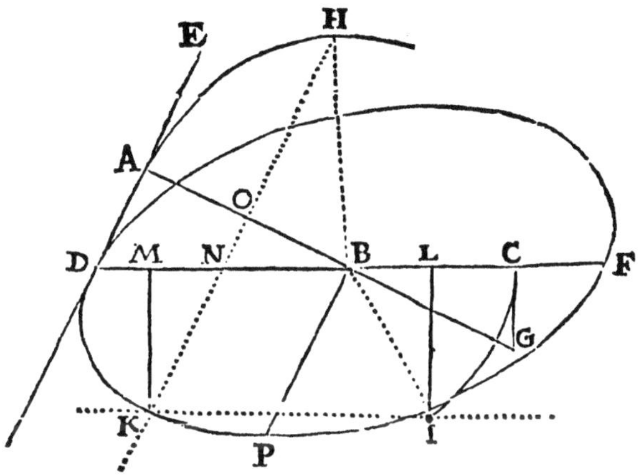

vam, continuâ ipſarum interſectione, veluti D vel K, deſcriptam, Ellipſin eſſe, cujus ſemi-diametri magnitudine ſunt rectæ D B, B G, nempe dictorum crurum, ſi opùs fuerit, productorum, portiones à perpendicularibus A D, C G, per aſſumpta puncta A & C reciprocè ductis, ad Polum interceptæ; & quidem altera, uti D B, etiam poſitione; altera verò, ut B G, non item, ſed B P ipſi æqualis, rectæque D A E æquidiſtans.

Sit enim prædictus A B C angulus in alia ſtatione utcunque, ex. gr., in H B I; ideoque præfata interſectio ad K. Demiſſis autem ab I & K ad rectam D F, ipſius D B duplam, perpendicularibus I L, K M, notatiſque interſectionem punctis ad N & O, quoniam æquales ſive iidem ſunt angulus A B C ſive O B L & H B I, erunt quoque, addito vel ablato communi H B F, anguli H B O & I B L æquales; ideoque triangula H B O & I B L, ob angulos præterea ad O & L rectos [1], æquiangula. Sunt autem [2] &

[1] ex hypotheſi, & per 29 primi.
[2] per 21 ſexti.

[241] the triangles CBG and MNK are also equiangular, as both are similar to the triangle OBN;[1] because[2] the square on DB is to the square on NB as the square on AB (i.e.,[3] the square on HB) is to the square on OB, therefore,[4] by conversion [4.12] the square on DB will be to the rectangle DNF[5] as the square on HB is to the square on HO,[6] i.e.,[7] as the square on BI is to the square on IL, or as the square on BC is to the square on KM,[8] i.e., as the square on BG or on BP is to the square on KN, and, by interchanging the means[10]: the square on DB is to the square on BP as the rectangle DNF is to the square on KN.

For this reason the curve $DKPF$, described by the intersection as we explained before,[11] is an ellipse with conjugate semidiameters DB and BP; so B is the center and DAE touches the ellipse at the vertex D.[12]

It should be pointed out here that, if the angle ABC should be a right angle, the point of intersection would describe not a curve but a straight line [4.12].

In the preceding (page [202], transl.) we described an ellipse by means of the motion of a point on one and the same straight line; now we have delineated it by the intersection of two straight lines.

---

[1] because the angles at C, O and M are right angles, but those at B and N are either identical or opposite angles . [2] VI, 4 and 22; [3] by hypothesis; [4] V, 19, Cor.; [5] II, 5; [6] I, 47; [7] VI, 4 and 22; [8] BC equals BI and IL equals KM; [9] VI, 4 because the triangles CBG and MKN are equiangular; [10] V, 16; [11] Prop. 13, p. [205]; [12] Prop. 13, Cor. 2, p. [209].

& æquiangula triangula C B G & M N K, cum tam hoc quàm illud triangulo O B N simile sit[1]: quare cum sit [2] D B quadratum ad N B quadratum, ut A B quadratum, id est [3], H B quadratum, ad O B quadratum, erit [4] per conversionem rationis D B quadratum ad D N F [5] rectangulum, sicut H B quadratum ad H O [6] quadratum, id est [7], uti B I quadratum ad I L quadratum, vel uti B C quadratum ad K M quadratum [8], id est [9], uti B G sive B P quadratum ad K N quadratum, & permutando [10] D B quadratum ad

[1] *ob angulos ad* C, O, *&* M *rectos, ad* B *verò &* N *sive eosdem sive ad verticem.*
[2] *per* 4 *&* 22 *sexti.* [3] *ex hypothesi.* [4] *per Cor.* 19 *quinti.* [5] *per* 5 *secundi.* [6] *per* 47 *primi.* [7] *per* 4 *&* 22 *sexti.* [8] *æqualis est enim* B C *ipsi* B I, *&* I L *ipsi* K M. [9] *per* 4 *sexti, propter triangula* C B G *&* M K N *æquiangula.* [10] *per* 16 *quinti.*

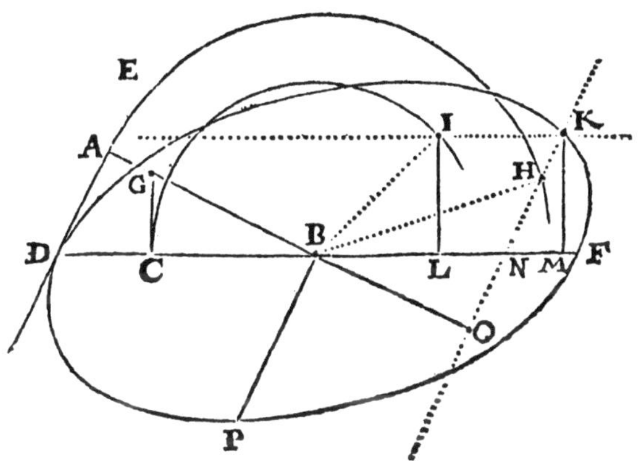

B P quadratum, ut D N F rectangulum ad K N quadratum. Ac proinde Ellipsis est curva D K P F, intersectione uti prædictum est descripta [11], cujus semi-diametri conjugatæ D B, B P; ideoque B centrum, ac D A E contingens Ellipsin in vertice D [12].

[11] *per* 13 *hujus.*
[12] *per* 2 *Cor.* 13 *hujus.*

Notandum hîc est, quòd si rectus foret A B C angulus, intersectione, uti prædictum est, non curvam, sed rectam lineam describi.

Quemadmodum autem Ellipsin, quæ superiùs per motum puncti in una eademque recta descripta fuit, nunc per duarum rectarum intersectionem delineavimus,

[242] In the same way a parabola and a hyperbola can also be described by means of the motion of a point on one and the same straight line, whereas in the preceding (pages [158] and [179], transl.) we explained their generation by similar intersections (i.e., by means of the intersection of two straight lines, transl.).

The ways of generating the curves we mentioned before are, however, infinite in number, as we already pointed out earlier, and we think we have explained the easiest and most natural methods.

Therefore it seems to us that we should not adhere to these topics any longer and so we proceed to finding and determining the plane and solid loci [4.13].

mus, ita & Parabola Hyperbolaque, quarum generationes ſolummodo per ſimiles interſectiones in præcedentibus expoſuimus, per motum puncti in una eademque recta deſcribi poſſunt. At verò quoniam prædictarum curvarum generationes, ut jam ante quoque monuimus, infinitæ ſunt, atque earum facillimas quidem ac maximè naturales à nobis jam propoſitas exiſtimamus, hiſce diutiùs inhærendum non videtur; itaque ad *Locorum Planorum*, *Solidorumque* inventiones ac determinationes progredimur.

IO-

4

Annotations to the translation

Chapter I

[1.1] Obviously, the other straight line is supposed to preserve its direction.

[1.2] The source text says "*angulus rectilineus*" (Euclid calls it γωνία εὐθύγραμμος; I def. 9). This means the angle between two straight lines, contrary to an angle between a straight line and a curve or between two curves. Of course, this is not necessarily a right angle. From now on, we shall simply translate "*angulus rectilineus*" by "angle."

Already in antiquity, the concept of "angle" raised extensive discussions. One question was whether an angle was a quantity, a quality, or a relation. Contrary to the view that an angle is a quantity, it was argued that the angle between a circle and a tangent at one of its points, a so-called horn-shaped angle (γωνία κερατοειδής, *angulus contingentiae*), does not satisfy the axiom of Archimedes-Eudoxus. In this case, the term "angle" is in the first place associated with an angle as part of the plane. For an extensive discussion, we refer to Lit. [22, 25].

[1.3] See Figure 4.1. For the description of a parabola, Jan de Witt assumes a straight line r (the directrix), a vertex T (the pole), and a fixed straight line m through T, intersecting the directrix at the point A; AT is called the interval or the parameter (p).

The curve in question is generated in the following way: an angle α (the moving angle or active angle) with arms s (the passive arm) and w (the active arm, the support of which is called the active line), rotates about the pole T, coinciding with its vertex. The rotating arm s intersects the

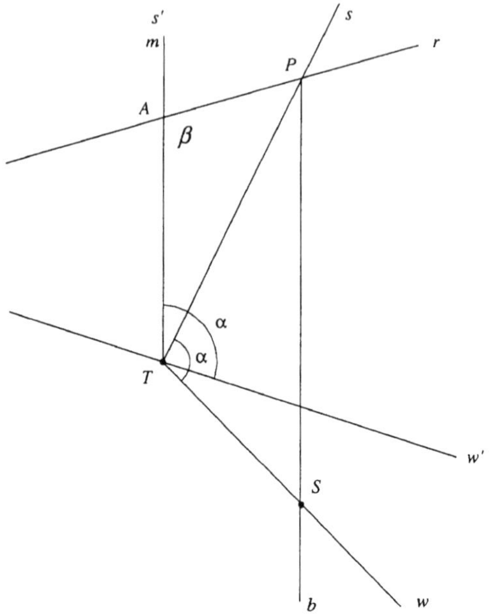

FIGURE 4.1

directrix r at a variable point P. Through P, a straight line b is drawn parallel to m; this line is called the generator. We now consider the locus of the point of intersection S of the generator b and the active arm w of the angle α.

This locus is a parabola if the angle α equals the angle β between m and r (at the same side of m).

A simple example can be found in Figure 4.2, where the angle α in question is a right angle. Here T is the origin of a rectangular system of axes, the x-axis of which coincides with m; r is the straight line with $x = -p$ as its equation. P is a variable point on r, the generator b is parallel to the x-axis, and, furthermore, the angle PTS (with S on b) is a right angle. Then, in the triangle PTS, we have $TN^2 = PN \cdot NS$; hence the coordinates x and y of $S(x, y)$, satisfy $y^2 = px$, which means that the point S generates a parabola, when the angle PTS rotates about the point T and when at the same time the point P drags the line b parallel to the x-axis.

If the angle β between m and r differs from α, a hyperbola is generated. This case is treated in Chapter IV.

[1.4] This is the case in which the generator passes through the pole (in the initial position, "*in statione prima*," to be abbreviated as "*i.s.p.*").

[1.5] In the translation, nevertheless, we will often add "i.s.p." (in statione prima) to leave no doubt.

[1.6] We will keep close to the Latin source and we will not translate "the square on the line segment AB" by AB^2. It may be remarked that "*recta*" means

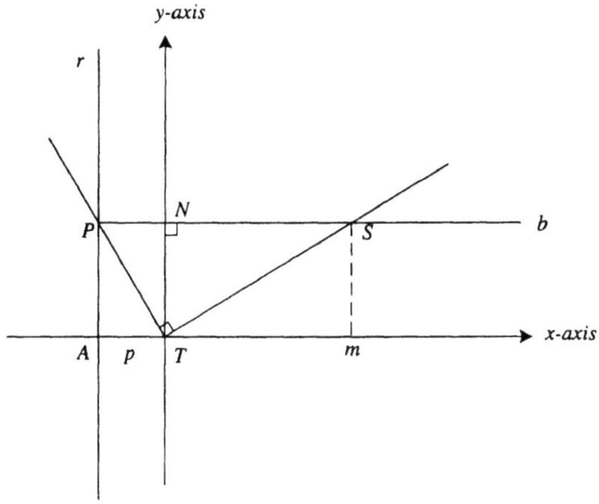

FIGURE 4.2

both "*straight line*" and "*line segment.*" This corresponds with the use of the term "γραμμή" in Euclid. There too, one often has to deduce from the context what is meant by this word. It is also interesting to remark that "$AB^2 = CD \cdot DF$" is represented in the Latin text by "*AB potest rectangulum CDF*," i.e.,: the line segment AB "is able" (i.e., after squaring) to produce the area of the rectangle with sides CD and DF, which means that the square with side AB has the same area as the rectangle with sides CD and DF.

Cf. the words *power*, *puissance*, and *Potenz.*'

[1.7] It concerns here D——B—K. By the rectangle DBK Jan de Witt means the product $DB \cdot BK$.

The proof given can be summarized as follows (see Figure III on page [165]):

$$\angle LDB = \angle LBD \text{ and } \angle LHI = \angle LIH,$$

hence $LD = LB$ and $LH = LI$,

which implies $LH - LD = LI - LB$, so $DH = BI$.

Furthermore, $\angle DBH = \angle IBG$

because $\angle DBI = \angle HBG$, so $\angle DBI - \angle HBI = \angle HBG - \angle HBI$.

Also $\angle BDH = \angle DBI = \angle BKG$, so $\Delta BDH \sim \Delta GKB$.

This implies $BD : DH = GK : KB$;

but $DH = BI = GK$,

hence $BD : GK = GK : KB$,

so $$GK^2 = DB \cdot KB.$$

[1.8] See Appendix B.

[1.9] The term "parameter" has been introduced by Claude Mydorge (1585–1647). For the term "ordinate-wise applied" (*ordinatim applicata*), see Appendix B.

[1.10] Here, "*linea*" (*recta*) clearly means "line segment." The term "diameter" is only now jusitified. In Theorem II, the existence of diameters will be proved.

[1.11] Clearly, a parabola and a straight line are supposed to have at most two points in common.

[1.12] $\angle ABD$ is the angle between the ordinate-wise applied chords and the diameter.

[1.13] It is typical of Jan de Witt that he already calls the parabola HAM, although the points H, A, and M only later get a well-defined meaning. See also Annotation [1.19].

[1.14] The third proportional to a and b is the quantity x that satisfies $a : b = b : x$. Here $AB : IM = IM : MK$ holds and hence, because $AB = IA$, we have $IM^2 = AI \cdot MK$.

[1.15] For the proof, see Figure III on page [171]. This can be summarized as follows: M lies on the original parabola HMA, hence, according to Theorem I, Proposition 1,

$$MB^2 = CA \cdot AB,$$

hence $$CA : 2MB = MB : 2AB.$$

But $AB = AI$, so $2AB = IB$; also $MB = GE$, hence

$$CA : 2GE = MB : IB.$$

But $\Delta IBM \sim \Delta DGH$, so

$$CA : 2GE = GH : GD,$$

and hence $$2GE \cdot GH = CA \cdot GD. \qquad (1)$$

Because H and M lie on the original parabola HMA:

$$HG^2 = CA \cdot AG \qquad (2)$$

and $$GE^2 = MB^2 = CA \cdot AB. \qquad (3)$$

(1), (2), and (3) imply:

$$(HG - GE)^2 = CA \cdot (AG - GD + AB) =$$

$$= CA \cdot (AD + AI) = CA \cdot ID = CA \cdot MO.$$

So $$HE^2 = CA \cdot MO. \tag{4}$$

Now K has been selected such that $MI^2 = AI \cdot MK$, while M lies on the parabola HMA, hence

$$MB^2 = CA \cdot AB.$$

Also, $AI = AB$, hence

$$BM^2 : MI^2 = CA : MK = CA \cdot MO : MK \cdot MO.$$

Combined with (4), this implies

$$BM^2 : MI^2 = HE^2 : MK \cdot MO.$$

Also, $$BM^2 : MI^2 = HE^2 : HO^2,$$

hence $$HO^2 = MK \cdot MO.$$

Therefore H lies on the parabola Π with SV as active line and KM as interval. H, however, has been arbitrary seleceted on the parabola HMA, which therefore coincides with the parabola Π mentioned above.

In connection with the sequel, the following can be remarked. In Figure III on page [171], a point P occurs, the role of which is not explained. From the context of the proof of Proposition 2, Corollary 5, on page [176], the following turns out to hold. If one extends MI to P such that $PI = IM$, then, from the construction of K (for $AI : IM = IM : MK$), it follows that

$$2AI : IM = 2IM : MK$$

and hence $$IB : IM = PM : MK.$$

Because $$\angle PIC = \angle MIB = \angle PMK,$$

$$\Delta IBM \sim \Delta MPK \text{ and also } \angle IBM = \angle MPK.$$

This fact is tacitly used in the proof of Proposition 2, Corollary 5, on page [176].

[1.16] In Figure III on page [175] this means for the tangent MI and the straight line MB, ordinate-wise applied to AG, that $BA = AI$ (A is the vertex), and conversely: if MB is ordinate-wise applied to AG and also if $BA = AI$, then IM touches the parabola at M.

[1.17] Four cases are distinguished here (see Figure III on page [175]):

i) The given point, M, through which one has to draw the tangent to the parabola, lies on the parabola and happens to be the vertex on a known diameter, while the direction of the straight lines, ordinate-wise applied to it, is known.
ii) The given point, M, lies on the parabola, but is not the vertex of a known diameter.
iii) The given point, I, lies outside the parabola, on the produced part of a known diameter.
iv) The given point, I, lies outside the parabola, but **not** on the produced part of a known diameter.

One preliminary remark: in Corollary 1 on pages [172] and [174], Jan de Witt already mentioned that one can easily find a diameter of a parabola by connecting the centers of two parallel chords; then one immediately has the straight lines ordinate-wise applied to this diameter. In this context, it must be pointed out that he thinks of a parabola specified as a curve drawn in the plane, and that one can intersect this curve simply with a straight line, as we intersect a circle with a straight line!
Now the four cases we distinguished:

i) It is evident that the tangent at the given point is parallel to the (known) straight lines that are ordinate-wise applied to the known diameter.
ii) See Figure III on page [175]. Here, the given point M is situated on the parabola. First, one constructs an arbitrary diameter AD with vertex A and corresponding ordinate-wise applied straight lines (like GH). Then one draws $MB \parallel GH$ (with B on AD), extends BA to I with the condition $AI = BA$, and then draws IM. Because of Corollary 2 on page [174], IM will touch the parabola at M.
iii) Here again, we look at Figure III on page [175]. In this case, I lies on a given diameter CG. We intersect this diameter with the parabola, say at A. Next we determine the point B such that $IA = AB$ and through B we apply a line ordinate-wise to AG. This is possible because, according to ii), we can construct the tangent to the parabola at A and hence we know the direction of the ordinate-wise applied straight lines. Let M be the point of intersection of this line and the parabola. According to Corollary 2, IM will then touch the parabola at M.
iv) If, however, I lies outside the parabola as well, but not on a given diameter, then we first determine a diameter together with the corresponding ordinate-wise applied straight lines. Next we draw a diameter IG through I and parallel to this diameter. Now we have the situation of case iii) and we can construct the tangent to the parabola from I in a similar way.

It must be pointed out, however, that Jan de Witt causes confusion, because in cases i) and ii), M is a given point on the parabola and in case iii) the point M turns up as the point of intersection of a certain line and the parabola.

[1.18] It concerns the following situation (Figure III, page [177]): a fixed diameter AG of a parabola is given, with vertex A, parameter AC, and given active angle. Besides, an arbitrary diameter MO is given. The proposition is that for the parameter MK belonging to MO the following proportion holds:

$$AI : IM = IM : MK.$$

Here M is the vertex of the diameter AG, and SMV is the tangent to the parabola at M.

[1.19] These are the angles that the straight lines that are ordinate-wise applied to MO and AB make with MO and AB, respectively. Incidentally, Jan de Witt is anticipating this with his notation. Thus, AB is already announced here as the diameter sought, prior to a proper definition of B as the intersection of CG and the line drawn through M under a certain angle with ME.

[1.20] Jan de Witt here uses the term "curves of the first genus" (*curvas primi generis*). Here he joins the classification of plane (algebraic) curves as given by Descartes in *La Géométrie* (1637; Livre Second, p. 319; see also the facsimile version in the Dover edition of 1954, p. 48). In fact, this is a classification according to the degree of the equation of the curve, in the coordinates x and y. The first genus (*premier genre*) contains the curves of which the equation has degree two or less. These are the straight lines, circles, parabolas, hyperbolas, and ellipses. The second genus contains curves with equation of degree three or four; the third genus contains curves with equations of degree five or six, etc. Of course, genus here does not have the modern meaning, which is defined as follows: by the genus of a plane algebraic curve of degree n, whose only singular points are d double points ($d \geq 0$), we mean the number p, defined by the formula

$$p = \frac{(n-1)(n-2)}{2} - d.$$

This peculiar classification is based on a misunderstanding by Descartes. He knew that one can solve a fourth degree equation by means of a third degree equation, the cubic resolvent, and he assumed–wrongly–that one could solve a sixth degree equation by means of a a fifth degree equation, etc. This explains the classification in classes of couples.

With this classification, Descartes opposes a classification into three genera, which occurs in the Συναγωγή (*Collectio*) of Pappus of Alexandria (early fourth century C.E.). Here the plane curves are classified in three genera, usually referrred to by the Latin terms *loci plani*, *loci solidi*, and *loci lineares*. We shall translate these terms by plane, solid, and linear loci.

Here "*locus*" stands for the set of points with a particular characteristic property.

The *loci plani* are the straight line and the circle. The *loci solidi* are the ellipse, parabola, and hyperbola, our conics to be precise (except the circle), because for their construction a three-dimensional object, a cone, is needed. These curves were defined by Apollonius of Perga (circa 250 B.C.E.) in the most literal sense as intersections of a cone with a plane (see also Appendix B). The other plane curves belong to the *loci lineares*, "because" –says Pappus–"for their construction other lines are needed than the ones I just mentioned, lines of deviant and more complex origin. Such lines are the spirals, the quadratrix, the conchoid, and the cissoid, which all have many important properties" (*Collectio*, vol. I, book III, Proposition 5). On page [228], Jan de Witt refers to this, when he speaks about "*distributionem in sua genera et species.*" Some used the term "mechanical curves" for curves that cannot be constructed by ruler and compass, like the quadratrix. Descartes, however, reserved this term for curves that cannot be represented by an algebraic equation; we call these curves transcendental –following Leibniz. In this case the cissoid and the conchoid do not belong to the mechanical curves, because they have the equations

$$x^3 = y^2(a-x) \text{ and } (x-a)^2(x^2+y^2) = b^2x^2.$$

Descartes called them "admissible" curves.

[1.21] Jan de Witt reasons as follows: till now he considered the case in which the moving angle equals the angle between the interval and the directrix (at the same side of the interval). It would now be most natural to treat the case in which the angles in question are different. He defers this, however, till Caput IV, page [201], and there the curve thus generated appears to be a hyperbola. Nonetheless, Jan de Witt first takes a different course for the following reason: in the first place, he is interested in the characteristic property of the hyperbola (*ea ipsius proprietas, quam primam ac maxime universalem existimo*), which he proves on page [181] as Proposition 3. Very briefly and incompletely stated, it is: $x \cdot y =$ constant. He can prove this property in the simplest way by means of a different way of generating the hyperbola, which he exposes in Caput II. This way does not involve a "rotating angle and a sliding line" but a "rotating line and a sliding angle."

Chapter II

[2.1] It is about the following situation (see Figure 4.3): for the description of a hyperbola Jan de Witt starts with a fixed straight line r (the directrix), a fixed point T (the pole) and a straight line b through T (the generator), which rotates about T and intersects the directrix r at a variable point A.

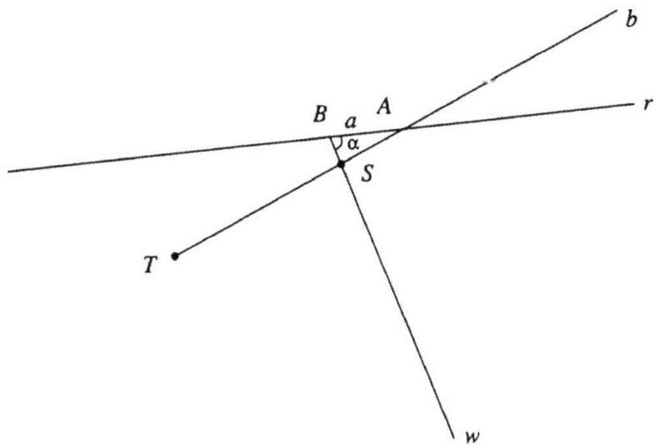

FIGURE 4.3

This point is the endpoint of the arm AB of fixed length a (the passive arm) of an angle with constant size α (the moving angle or active angle). The arm AB always lies on the directrix r and is pushed along this directrix when b rotates about the pole T.

We now consider the locus of the point of intersection S of the rotating generator and the other arm w (the active arm) of the angle α. It is evident that the arm w always preserves a fixed direction.

When the generator is parallel to the active line (i.e., the support of the active arm), both are said to be in the initial position ("*in statione prima*," often abbreviated as i.s.p.). The part TA' of the generator in the initial position that lies between T and the directrix r is called the interval. This name is also reserved for the length a of the passive arm (see Figure 4.4). It turns out that the point S generates a hyperbola whose asymptotes

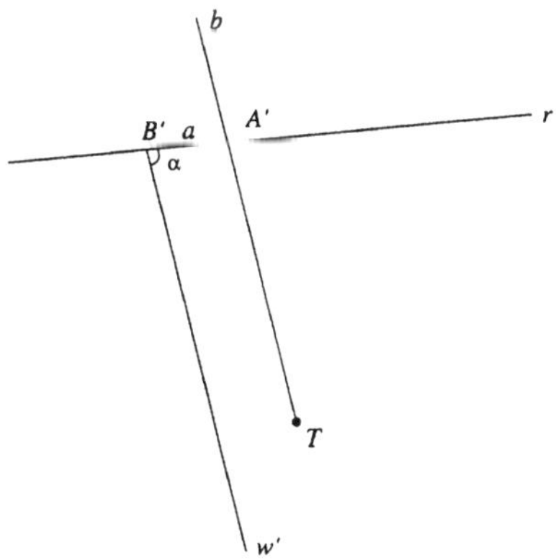

FIGURE 4.4

are the directrix r and the active line w' in the initial position. Of course, the point of intersection B' of these two lines is the center.

Either branch is considered by Jan de Witt as a single curve. In this case, he speaks of opposite hyperbolas (*hyperbolae oppositae*) and he calls each branch a hyperbola.

For the proof that the curve, constructed in this way, is indeed a hyperbola, we first look at Figure 4.5; in a subsequent note, we will pay attention to the figure on page [181]. In Figure 4.5, S is an arbitrary point

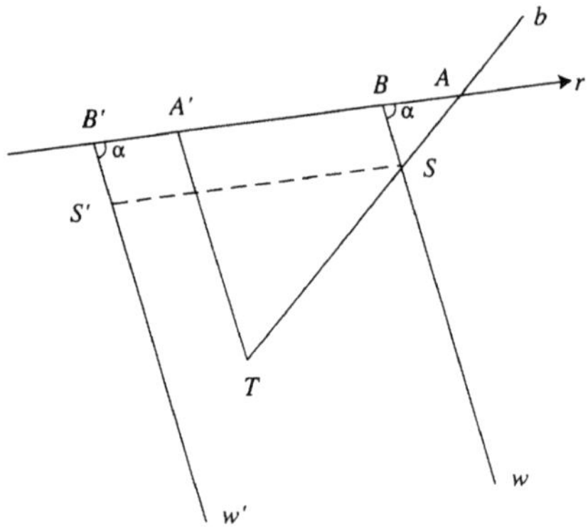

FIGURE 4.5

on the curve; the generator TA' and the active line w' are in the initial position, which means: $TA' \parallel w'$ and hence we have $AA' = BB'$, because $AB = A'B'$. If we also draw SS' parallel to the directrix r, then we have in $\Delta AA'T$

$$AB : BS = AA' : A'T,$$

so

$$AB : BS = BB' : A'T,$$

hence

$$BS \cdot SS' = BS \cdot BB' = AB \cdot A'T = \text{constant}.$$

We now select an oblique system of axes with B' as its origin and r and w', in the prescribed direction, as positive x-axis and y-axis, respectively. The equation of the curve then reads

$$x \cdot y = c.$$

Here c is the constant product of both intervals; c is called the power of the hyperbola. In this way, the content of Theorem III on pages [180] and [181] is given.

[2.2] The proof on page [182] is based on the figure on page [181]. At first sight, this may cause confusion, because three different points are there denoted by the letter C, three by the letter B, and three by the letter E. It is, however, evident that this concerns the generator ACB and the angle BEC

in three different positions; a fourth position of this line and this angle occurs as Abc, and bcc, respectively. Jan de Witt points to this fact in the corresponding marginal note a. If one starts from the line ACB in the lowest position, one quickly recognizes the situation of Annotation [2.1] and Figure 4.5, and one can easily follow the reasoning here as well.

[2.3] By "the rectangle FEC," Jan de Witt means the area of the rectangle with sides FE and EC, i.e., the product $FE \cdot EC$.

[2.4] It is evident that in this case too, the line *segments* EB, FD, FB, ED, BD, and FE are designated by the term "straight line" (*linea*). See also Annotation [1.6].

[2.5] I.e., they have the same area.

[2.6] See introductory letter.

[2.7] The source text gives C; this must be c. Apparently, NP has been selected to be parallel to ec and GFH.

[2.8] It concerns the asymptote GFH.

[2.9] It concerns an intersecting line and a tangent.

[2.10] This refers to $FBCG$. See also Corollaries 2 and 3 on page [184].

[2.11] This refers to MCL.

[2.12] We would say: on one branch or on two different branches of a hyperbola.

[2.13] In Figure I on page [185], this means: $EB \cdot BF = GC \cdot CH$ and $EB \cdot BF = FD \cdot DE$, while $GC \cdot CH = HP \cdot PG$. In Figure II on page [187], $EB \cdot BF = FD \cdot DE$.

[2.14] It is also evident that in this context the two branches of the hyperbola are considered as two distinct curves.

[2.15] In Figure III on page [187] the relations $EB \cdot BF = GP^2$ and $FD \cdot DE = CG^2$ hold.

[2.16] When one switches from the ratio $a : b$ to the ratio $(a+b) : b$, one says that the second ratio originates from the first one by composition (*compositio*, *componendo*, σύνθεσις). When one switches from the ratio $a : b$ to the ratio $(a-b) : b$, one says that the second ratio originates from the first one by separation (*separatio*, *separando*, διαίρεσις). Instead of *separando*, sometimes *dividendo* is used.

In Book V of *Elementa* by Euclid, Theorems 17 and 18, we find the following statements:
If magnitudes be proportional *componendo*, they will also be proportional *separando* (Theorem 17) and: if magnitudes be proportional *separando*, they will also be proportional *componendo* (Theorem 18). This means, respectively:

$$(a+b) : b = (c+d) : d \Rightarrow a : b = c : d$$

and

$$(a-b) : b = (c-d) : d \Rightarrow a : b = c : d.$$

[2.17] By way of illustration: in Figure I on page [189], the half line ANO that intersects the hyperbola is an intercepted diameter, and if $PN = NC$,

PNC is said to be applied ordinate-wise to ANO. This holds for all chords parallel to PNC. A straight line through A that intersects neither branch of the hyperbola is called a second diameter. If such a diameter is parallel to the chord PNC just defined, then this second diameter is said to be conjugate to the diameter ANO. The relation "conjugate to" will appear to be symmetric. On page [192] a length will be assigned to a second diameter.

[2.18] See Figure 1 on page [189]. This can be summarized as follows: $OD = OB$ and $DF = BE$, hence $OF = OE$. Now suppose that $OR = OQ$; this would imply $OS = OT$, because $RS = QT$. If one draws $EV \parallel SF$, then certainly $OV < OT$, but $EO : OV = FO : OS$ and $EO = FO$; this would imply $OV = OS = OT$, which leads to a contradiction.

[2.19] So in the figure on page [191] we have:

$$EB \cdot BF = BF \cdot FD =$$

$$= FD \cdot DE = DE \cdot EB = CH^2 = CG^2.$$

[2.20] I.e., the point of intersection of this diameter and the hyperbola.

[2.21] See also Annotation [2.29].

[2.22] First this (see the figure on page [193]): the second diameter AK is ordinate-wise applied to the transverse diameter EC, if it is parallel to the tangents at the vertices E and C. By definition, a straight line HI is again ordinate-wise applied to the second diameter AK, if the chord HI is bisected by AK. The proposition says that in this case HI is parallel to the original transverse diameter EC.

In proving this, Jan de Witt does not a priori start with a diameter that is ordinate-wise applied to AK, but with a straight line HI determined by the property that it passes through the points G and D, where the tangents to the hyperbola at E and C intersect the asymptotes. This line intersects the hyperbola at H and I. He proves two things about this line HI :

1. This line is parallel to EC;
2. $KH = KI$, hence HI is ordinate-wise applied to AK.

This last property does not appear before line 4 from the bottom on page [193]: "*Quocirca cum ad secundum diametrum AK applicata sit recta HI...*," where addition of the term "*ergo*" after "*cum*" might have given some clarification.

The theorem then follows from the facts that only one straight line ordinate-wise applied to AK passes through K (Theorem V, Corollary 6), and that all lines ordinate-wise applied to AK are parallel to each other.

[2.23] These diameters are given in position and in size, i.e., if C is the point at which the given diameter intersects the hyperbola and if A is the center (point of intersection of the asymptotes), then both CP $(= 2AC)$ and the length of the part GH of the tangent to the hyperbola at C between the asymptotes (see the figure on page [194]) are given.

[2.24] The proof can be summarized as follows: C lies on the hyperbola, hence $CB \cdot BA$ is the characteristic constant for this curve (also known as the power, as on page [182]). In accordance with the construction of D and E, $AB : AD = AD : BC$, hence $AD^2 = AB \cdot BC$; however, $AD = DE$, hence $ED \cdot DA = CB \cdot BA$ (the power of the hyperbola). Also, $DE \parallel AK$, hence E lies on the hyperbola. It can now simply be recognized that $IE = EK$ and that $IE \perp AE$. This implies that FAE and IEK are the conjugate axes of the hyperbola.

[2.25] See the figure on page [195]. Here AG and AK are the asymptotes of a hyperbola, on which the points C and E have been selected, with tangents GCH and IEK. Five propositions are stated:

$$
\begin{array}{lrcl}
1) & \text{area } \Delta GAH & = & \text{area } \Delta KAI. \\
2) & GA \cdot AH & = & KA \cdot AI. \\
3) & GI : IA & = & KH : HA. \\
4) & GR : RH & = & KR : RI. \\
5) & GC : CR & = & KE : ER.
\end{array}
$$

In the source text, 2) is proved first, and only then is 1) proved. In the sequel, we will preserve this order and prove 2) first ("*quod est primum*").

2) Draw CB and $ED \parallel AK$, then $GC : GH = GB : GA = BC : AH$, but $GH = 2GC$, hence $GA = 2GB$ and $AH = 2BC$,

so $$GA \cdot AH = 4GB \cdot BC = 4AB \cdot BC \qquad (*)$$

and $$KA \cdot AI = 4AD \cdot DE. \qquad (**)$$

Because the right-hand sides in (*) and (**) represent four times the power of the hyperbola, we conclude

$$GA \cdot AH = KA \cdot AI. \qquad (2)$$

1) For the proof of 1), Jan de Witt appeals to *Elementa* VI, 15: if two triangles have an angle in common and if the sides containing this angle are reciprocally proportional, then these triangles have the same area. According to 2), this condition is satisfied.
3) From 2), we have $GA : IA = AK : AH$. By separation (*separando* or *dividendo*, see Annotation [2.16]), we can prove that

$$(GA - IA) : IA = (AK - AH) : AH$$

and hence $GI : IA = KH : AH$, which is 3).

4) Area ΔGAH = area ΔKAI (see 1)). If one omits the common area $AIRH$ in both triangles, then area ΔGRI = area ΔKRH, with equal angles at R. According to *Elementa* VI, 15, the equation

$$GR : RH = KR : RI \quad \text{applies, which is 4).}$$

5) From the preceding proportion, we can derive by composition (Annotation [2.16])

$$(GR + RH) : RH = (KR + RI) : RI,$$

hence $\quad GH : RH = KI : RI;$

however, $\quad GH = 2CH$ and $KI = 2EI,$

so $\quad CH : RH = EI : IR.$

Now, the concept of "conversion" (*conversio*, ἀναστροφή) is used. This implies (*Elementa* V, Definition 16)

$$a : b = c : d \Rightarrow a : (a - b) = c : (c - d),$$

so $\quad CH : CR = EI : ER$

and hence $\quad GC : CR = KE : ER$ and this is 5).

[2:26] See the figure on page [198]. Here, $PACN$ is the support of the arbitrary diameter under consideration, GH is the corresponding second diameter, and PAC is the transverse diameter (the last two considered as line segments). BND is an aribitrary chord that is ordinate-wise applied to ACN, i.e., $BN = ND$.

By "those segments of the same diameter that are cut off by either end of the transverse diameter (i.e., P and C, translator) and the line applied," Jan de Witt means the line segments with endpoints P, C, and N. In itself, this is not unambiguous. From the proof, it is evident that it concerns the line segments PN and CN, hence the theorem implies: $GH^2 : CP^2 = DN^2 : PN \cdot CN$.

The proof is as follows:

Due to Proposition 5 on page [185],

$$HC^2 = BF \cdot FD,$$

hence, because $\Delta ANF \sim \Delta ACH$:

$$FN^2 : HC^2 = FN^2 : BF \cdot FD = NA^2 : CA^2, \qquad (*)$$

hence $\quad (FN^2 - BF \cdot FD) : HC^2 = (NA^2 - CA^2) : CA^2$

and $$[(ND + FD)^2 - (2ND + FD) \cdot FD] : HC^2 =$$

$$= (NA + CA) \cdot (NA - CA) : CA^2,$$

so $$ND^2 : HC^2 = PN \cdot NC : CA^2,$$

which implies $$ND^2 : PN \cdot NC = HC^2 : CA^2 = GH^2 : CP^2.$$

The parameter CI (see page [192]) is defined by:

$$CP : GH = GH : CI \Rightarrow GH^2 = CP \cdot CI,$$

so $$GH^2 : CP^2 = CI : CP = DN^2 : PN \cdot NC.$$

[2.27] The center is defined as the point of intersection of two diameters, and a diameter is defined as the line connecting the midpoints of two parallel chords. For the latter definition it is necessary that one determines the points of intersection of a line and the hyperbola. This is apparently done very naively: the hyperbola is drawn as a curve on the paper and is simply intersected by a straight line. There is no question of a ruler and compass construction of the point of intersection of a line and a hyperbola defined by five independent points or by equivalent data.

[2.28] By "the rectangle CNK" the product $CN \cdot NK$ is meant–as always. The proof reads in modern formulation: according to Proposition 10 (see also Annotation [2.26]):

$$PC : CI = PN \cdot NC : DN^2. \qquad (*)$$

From the figure on page [198], it appears that

$$PC : CI = PN : NK = PN \cdot NC : NK \cdot NC\ . \qquad (**)$$

From (*) and (**), it is immediately evident that $DN^2 = CN \cdot NK$.

[2.29] For an explanation, the reader is referred to Figure 4.6. This is in fact the figure on page [198], with some additions.

Here BCD is the hyperbola referred to. Its asymptotes are AG and AH, while PAC is a diameter; GCH touches the hyperbola at C and hence has the same length as the second diameter conjugate to PAC. The point I lies on the produced part of GCH, such that CI is the latus rectum (parameter) belonging to the selected diameters, which means that $CI = GH^2/PC$. Through the arbitrary point D on the hyperbola, a chord DNB has been ordinate-wise applied to PAC, hence $DNB \parallel HCG$. Hence, the directions of PAC and DNB are mutually conjugate (see Appendix B). Furthermore: $PR \parallel GCH$ and $RIS \parallel KT \parallel PAC$.

In Corollary 2, Jan de Witt proved: $DN^2 = CN \cdot NK$, but now he interprets this result –somewhat cryptically–"*veterum Geometrarum more,*"

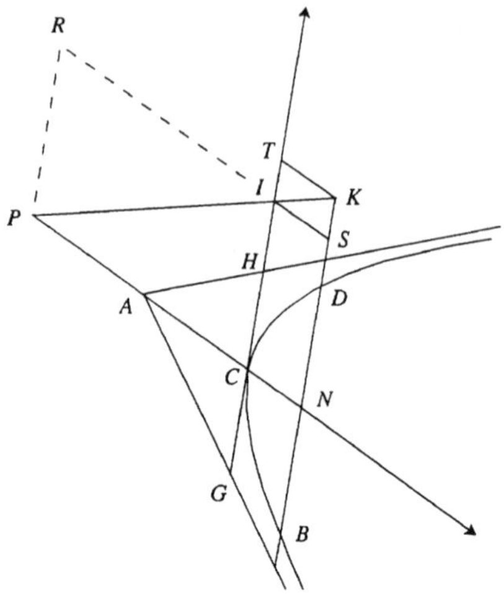

FIGURE 4.6

i.e., in the fashion of the "classical geometers." For a good understanding of this, we consider Figure 4.6 of this annotation.
We then see

$$DN^2 = CN \cdot NK = CN \cdot (NS + SK) =$$

$$= CN \cdot NS + CN \cdot SK = CN \cdot CI + SI \cdot SK.$$

DN is ordinate-wise applied to PAC; its square turns out to be equal to the area of the rectangle whose sides are CI (the parameter or latus rectum) and CN (that part of the diameter intercepted between the vertex and the applied line DN), augmented by the area of the rectangle with SI and SK as its sides.

The rectangle $SKTI$ is similar to $CIRP$, and $CIRP$ has as its sides the parameter CI and the transverse diameter PC. Moreover, $SKTI$ has "the same position" as $CIRP$, which means that their corresponding sides are parallel. For the background of this way of formulating by the "classical geometers," we refer to the treatment of the hyperbola in Appendix B.

[2.30] In the figure on page [199] this means that, if ECF is the arbitrary tangent and if AH is the arbitrary diameter, then $AI \cdot AH = KA^2$, hence

$$AH : KA = KA : AI.$$

It may be remarked that KA is half a diameter and hence the line DKG, which is parallel to the line CH ordinate-wise applied to AH, touches the hyperbola at K. The proof can be summarized as follows:

if $\qquad DKG \parallel CH$ and $KL \parallel ECF$,

then according to Proposition 9, sub 5, page [195],

$$RC : CF = RK : KD$$

so $$MG : MF = LE : LD$$

and also $$MG : GF = LE : ED. \qquad (*)$$

Furthermore, $$GF : GA = ED : EA \qquad (**)$$

(Proposition 9, sub 3, page [195]), hence (*) and (**) imply

$$MG : GA = LE : EA.$$

Here Jan de Witt appeals to *Elementa* V, 22 and translates the term δι' ἴσου (*di isou*) by the term *ex aequo*. One sometimes finds the term *per aequale* for it, but usually *ex aequo*.
This means the following: if a_1, a_2, a_3, and b_1, b_2, b_3 are two sequences of quantities for which

$$a_1 : a_2 = b_1 : b_2 \text{ and } a_2 : a_3 = b_2 : b_3,$$

then one can easily verify that also $a_1 : a_3 = b_1 : b_3$.
In our case, it concerns the sequences MG, GF, GA and LE, ED, EA.
One says that this last ratio holds *ex aequo*, because the terms on the left- and right-hand sides of these ratios have corresponding places in the sequences, hence they have the same distance (*distantia*).
Generally, if two sequences of quantities $a_1, a_2, \ldots, a_n$ and $b_1, b_2, \ldots, b_n$ are given with the relations

$$a_i : a_{i+1} = b_i : b_{i+1} (1 \leq i < n),$$

then *ex aequo*

$$a_1 : a_n = b_1 : b_n.$$

As to the terms "ratio" and "quantity," one can think of Definition III in *Elementa* V: "Ratio is a certain relation in size between two similar quantities" and Definition IV of the same book: "Those quantities that can outnumber each other after multiplication, are said to be able to have a ratio."
The proof continues as follows:

From $$MG : GA = LE : EA,$$

it follows that $$HK : KA = MG : GA = LE : EA = KI : IA$$

and hence $HA : KA = KA : IA.$

[2.31] In the figure on page [200], the diameters AKH and QAB are conjugate. The line FCQ touches the hyperbola at C; the tangent at the vertex K to the hyperbola intersects the asymptotes at D and G; furthermore: $BC \parallel HA$.

The proposition is that $KG^2 = BA \cdot AQ$. Its proof runs as follows:
From the figure it is evident that

$$HA^2 : KA^2 = HM^2 : KG^2.$$

According to Proposition 11,

$$KA^2 = HA \cdot IA,$$

hence $HA^2 : KA^2 = HA : IA.$

In the figure one immediately sees:

$$HA : IA = CB : IA = BQ : AQ,$$

so $HM^2 : KG^2 = BQ : AQ.$

Separatio (see Annotation [2.16]) gives:

$$(HM^2 - KG^2)^2 : KG^2 = (BQ - AQ) : AQ.$$

According to Proposition 9, $KG^2 = TM \cdot MC$; by this equality and by the figure on page [200], we find:

$$\{(HC + CM)^2 - (2HC + CM) \cdot MC\} : KG^2 = BA : AQ,$$

so $HC^2 : KG^2 = BA : AQ$

and, because $HC = AB$,

$$AB^2 : KG^2 = BA : AQ.$$

Jan de Witt interprets this geometrically, referring to *Elementa* VI, 20, where it is stated that the areas of similar figures (here squares with sides AB and KG) have the same ratio as the squares of corresponding sides. Here this means

$$KG^2 = (AQ/AB) \cdot AB^2$$

and hence $KG^2 = AQ \cdot AB.$

[2.32] For illustration of the reasoning on page [201], see the figure on page [200].
[2.33] According to Proposition 6 on page [190].
[2.34] According to Theorem X on page [199].
[2.35] According to Theorem XI on page [200]. Incidentally, in line 7 from the bottom one must read AB instead of AH.
[2.36] According to Theorem III, Corollary 1 on page [182].

Chapter III

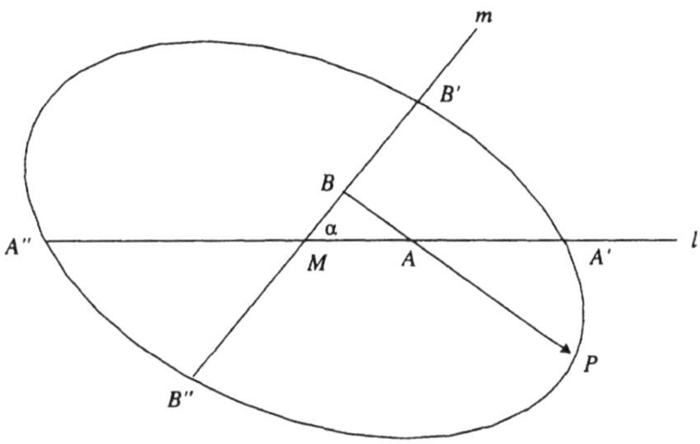

FIGURE 4.7

[3.1] Without prior notice, Jan de Witt also considers the case in which the angle CAB is not a right angle (page [204], line 1 from the bottom and page [205], line 1 from the top). It concerns the following general case (see Figure 4.7):
Let an angle α be given with vertex M and arms l and m. A line segment AB with fixed length has one end, say A, on l and the other end on m. On AB, or on the produced part of AB or of BA, one selects a point P. The issue is the locus of the point P, if AB is moving such that A always remains on l or on its produced part, and B always remains on m or on its produced part.

Here the following terms are used: M is the center, the angle α is the active angle or simply the angle, AB is the generator; P is called the active point or simply the point, PA and PB are called the intervals.

When one measures the interval PB along l on either side of M, with endpoints A' and A'', then the line segment $A'A''$ is called the directrix. Similarly, one can measure the interval PA along m on either side of M with endpoints B' and B''; $B'B''$ is then called the directrix too. The generator AB is said to be in the initial position if it is perpendicular to the directrix $A'A''$ (see Figure 4.8). Then the active point P is in the initial position as well. The double of MP in that position is called the secant; in Figure 4.8, this is PP'. Similar definitions can be given, starting from the directrix

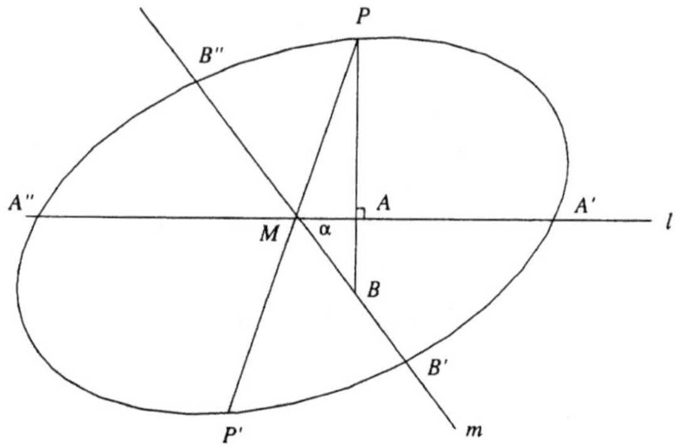

FIGURE 4.8

$B'B''$; of course, this yields a different secant.

[3.2] It is not hard to verify analytically that the locus of P is an ellipse. To this end, we look at Figure 4.9 and put $PB = a$, $PA = b$ and $P = P(x, y)$. In $\Delta\ PAD$, we then have:

$$\frac{b}{\sin\alpha} = \frac{y}{\sin\lambda},$$

hence

$$\sin\lambda = \frac{y}{b}\sin\alpha. \qquad (*)$$

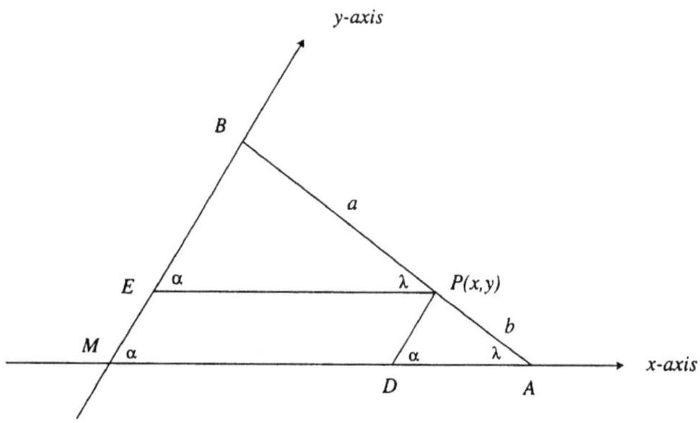

FIGURE 4.9

In $\Delta\ BPE$ we have:

$$\frac{a}{\sin\alpha} = \frac{x}{\sin(\alpha+\lambda)},$$

hence $$a(\sin\alpha\cos\lambda + \cos\alpha\sin\lambda) = x\sin\alpha.$$

Combined with (*), this yields

$$\cos\lambda = \frac{x}{a} - \frac{y}{b}\cos\alpha \quad (\text{if}\sin\alpha \neq 0). \tag{**}$$

The case $\sin\alpha = 0$ is trivial.

Combined with $\sin^2\lambda + \cos^2\lambda = 1$, (*) and (**) result in

$$\frac{y^2}{b^2}\sin^2\alpha + \frac{x^2}{a^2} - 2\frac{xy}{ab}\cos\alpha + \frac{y^2}{b^2}\cos^2\alpha = 1,$$

hence $$\frac{x^2}{a^2} - \frac{2xy\cos\alpha}{ab} + \frac{y^2}{b^2} = 1.$$

This is the equation of an ellipse, because the discriminant satisfies the relation

$$\frac{\cos^2\alpha - 1}{a^2b^2} < 0, \quad \text{if} \quad \alpha \neq 0, \pi.$$

[3.3] The correctness of this theorem can simply be shown analytically, if we

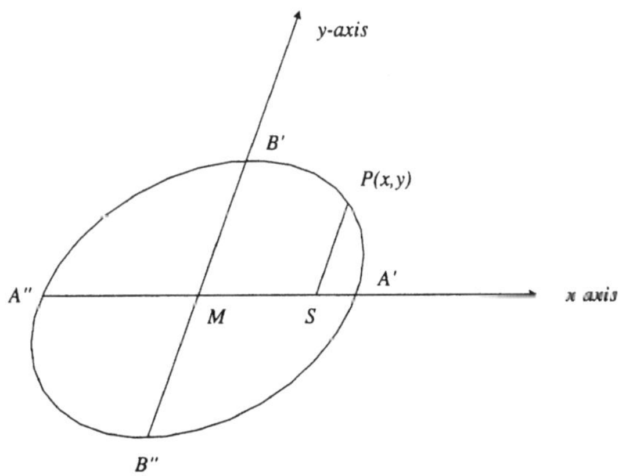

FIGURE 4.10

start from the equation of an ellipse with two conjugate diameters, e.g., $A'A''$ and $B'B''$, as coordinate axes. See Figure 4.10; here we put

$$MA' = a, MB' = b \text{ and } PS \parallel MB'.$$

We then have

$$\frac{x^2}{a^2}+\frac{y^2}{b^2}=1 \Rightarrow b^2x^2+a^2y^2=a^2b^2 \Rightarrow b^2(a^2-x^2)=a^2y^2,$$

hence $$\frac{y^2}{(a-x)(a+x)}=\frac{b^2}{a^2}$$

and $$PS^2 : A''S \cdot SA' = (B'B'')^2 \cdot (A'A'')^2.$$

[3.4] FAG in Figures I and II on page [206]; HAG in the other figures.

[3.5] To this case, we look at Figure I and Figure II on page [206].
In Figure I, we have:

$$KL = AE = AD, \text{ hence } KL^2 = AE^2$$

and $$KL^2 - KN^2 = AE^2 - AI^2,$$

hence $$LN^2 = (AE + AI)(AE - AI) = DI \cdot IE.$$

Also, $$LI^2 : LN^2 = LM^2 : LK^2 = FA^2 : AE^2,$$

so $$LI^2 : DI \cdot IE = FA^2 : AE^2 = FG^2 : DE^2.$$

[3.6] To this case, Figures III–VIII on pages [206]–[208] apply. We confine ourselves to Figure III, the other cases are treated in a similar way. Figure III is constructed as follows: BC is the generator in the initial position, i.e., perpendicular to DAE. The point L is an arbitrary point on the ellipse, LI is parallel to the secant HAG; KML is the generator for the point L; KO and LP are both perpendicular to AE. The line through I, parallel to AB, intersects LP at N; finally, KN is drawn.
The first aim is to show that $KN = AI$. This is done as follows:

$$BA : KA = BC : KO = MK : KO =$$

$$ML : LP = HC : LP = HA : LI = BA : NI,$$

i.e., $$BA : KA = BA : NI, \text{ hence } KA = NI.$$

Also, $$KA \parallel NI,$$

hence $$KN = AI \text{ and } KN \parallel AI.$$

Furthermore, $$KL = AE = AD.$$

We already saw $KN^2 = AI^2$,

hence $$LN^2 = KL^2 - KN^2 = AD^2 - AI^2 = DI \cdot IE.$$

Also, $$LI^2 : LN^2 = AH^2 : BH^2 = AH^2 : AE^2,$$

hence $$LI^2 : DI \cdot IE = AH^2 : AE^2 = HG^2 : DE^2.$$

[3.7] For clarity's sake, the first paragraph on page [209], from "*ita ut*" until "*secatur*" has been translated in the singular form.
In Theorem XIII, Proposition 14 on page [213], it will be shown that for every transverse diameter there does exist another diameter that satisfies the above definition of second diameter. Furthermore, it may be remarked that the diameters that are conjugate in this sense also have the characteristic property that is to be satisfied by conjugate diameters, i.e., all chords parallel to one are bisected by the other. This will be proved in Corollary 4 on page [211]. There, the term *ordinatim applicata*, i.e., ordinate-wise applied, is used without explanation. The meaning, however, is evident, because of the parallelism with the hyperbola. Furthermore, we refer to the treatment of the ellipse by Apollonius, as outlined in Appendix B.

[3.8] It is clear that the size of the parameter depends on the choice of the mutually conjugate diameters and their order. In modern notation: if the equation of the ellipse with respect to coordinate axes coinciding with conjugate diameters reads:

$$\frac{x^2}{a^2} + \frac{y^2}{b^2} = 1,$$

and if one locates the transverse diameter on the x-axis and the second diameter on the y-axis, then the corresponding parameter p is defined by $2a : 2b = 2b : p$, hence we have:

$$p = \frac{2b^2}{a},$$

but one can also interchange the role of a and b; then the corresponding parameter is $2a^2/b$.

For the concept of parameter, see also Appendices A and B. In the current literature, one mainly uses as parameter half the parameter introduced here.

[3.9] It must be emphatically mentioned here that the symmetry in the reasoning of the first part of the proof of Theorem XII is only valid because it concerns mutually perpendicular axes, as is evident from page [205], line 3 from the bottom. For the case of an arbitray diameter and its conjugate, see Annotation [3.15].

[3.10] In Figure I on page [210], it is immediately evident that $WX^2 : DX \cdot XE = FG^2 : DE^2 = LI^2 : DI \cdot IE$, by Theorem XII.
If one considers DE as a transverse diameter and FG as the corresponding second diameter, then the corresponding parameter p satisfies:

$$DE : FG = FG : p,$$

hence also $WX^2 : DX \cdot XE = FG^2 : DE^2 = p : DE$

and of course $WX^2 : LI^2 = DX \cdot XE : DI \cdot IE.$

The same applies to the situation in Figure II, although WX has not been drawn there.

[3.11] The term *ordinatim applicata* (ordinate-wise applied) has not yet been defined for an ellipse. The intention, however, is clear: a chord is called ordinate-wise applied to a diameter if this chord is parallel to the corresponding conjugate diameter.

The concept of conjugate diameter is defined at the end of Theorem XII. Till then, the only examples are the axes and the directrix and secant in the construction of an ellipse. Not until Theorem XIII, on page [213], is it shown that every diameter has a corresponding conjugate diameter.

[3.12] So, for the time being, it is only shown that very specific straight lines have at most two points in common with an ellipse, i.e., lines that have been ordinate-wise applied to a diameter. As remarked before, it will be proved in Theorem XIII that every straight line is ordinate-wise applied to some diameter, which proves that an arbitrary straight line has at most two points in common with an ellipse.

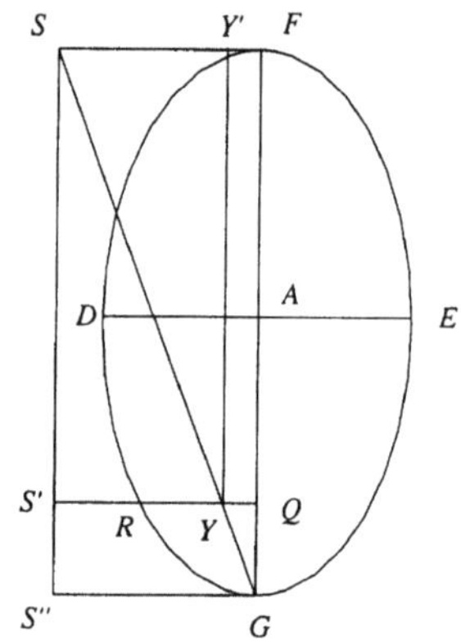

FIGURE 4.11

[3.13] See Figure I on page [212] and Figure 4.11 added in this annotation. In Figure I, GF has been selected as a transverse diameter (here even as an axis) and DE has been selected as a second diameter. According to the definition of parameter on page [209], the corresponding parameter p ($= FS$) satisfies $GF : DE = DE : p$ and hence

$$GF^2 : DE^2 = GF^2 : GF \cdot FS = GF : FS =$$

$$= GQ : YQ = GQ \cdot QF : YQ \cdot QF,$$

so $$GF^2 : DE^2 = GQ \cdot QF : YQ \cdot QF. \qquad (*)$$

According to Theorem XII,

$$GF^2 : DE^2 = GQ \cdot QF : RQ^2. \qquad (**)$$

From (*) and (**) it follows that:

$$RQ^2 = YQ \cdot QF.$$

The link with the practice of the classical geometers (see also Appendix B) does require some clarification, which is given here. Here, we refer to Figure 4.11, which is a somewhat extended version of Figure I on page [212]. In our Figure, we have:

$$RQ^2 = YQ \cdot QF = (S'Q - S'Y) \cdot QF =$$

$$SF \cdot QF - S'Y \cdot S'S = SF \cdot QF - Y'Y \cdot YS'.$$

Here RQ^2 is the square on the line segment RQ that has been ordinate-wise applied to the diameter FG; $SFQS'$ is the rectangle adjacent to the latus rectum (parameter) SF, with as width the line segment FQ, intercepted between the endpoint F of the diameter FG and the point Q; the construction of Q is evident from Figure 4.11. The rectangle $Y'YS'S$ is similar to the rectangle $FGS''S$, which is contained by the transverse diameter FG and the parameter SF. This rectangle $Y'YS'S$ has, moreover, the same position as the rectangle contained by the transverse diameter FG and the parameter SF, which means that the corresponding sides of $Y'YS'S$ and $FGS''S$ are parallel. This is the background to the final sentence of Corollary 6 on page [213]; the square RQ^2 is not equal to the area of the rectangle $SFQS'$, but the area of the rectangle $SFQS'$ is lacking (ἔλλειψις, see also Appendices A and B).

[3.14] Of course, Jan de Witt means $AB - AF$.

[3.15] The reasoning in the proof of Proposition 14 is as follows. We confine ourselves to Figure I on page [214] (= Figure 4.12), the other cases are similar. The reader is referred to the accompanying marginal notes by Jan de Witt. The ellipse $SYXZ$ is given by the axes SAX and YAZ. In this ellipse one draws a diameter DAE and looks for the corresponding conjugate (= second) diameter or the secant (for the definition, see page [205]). In order to determine this, a new ellipse is constructed in the following way: First, one defines a point H on $SYXZ$–in a way to be specified later –and draws HCB perpendicular to DAE such that $HB = DA$. Next, one considers the ellipse E^* with BAE as its angle and HC and HB as the intervals (so BC is the generator). According to the construction, BCH is the generator of this ellipse in the initial position, and hence (since $HA = AG$) HAG is the secant or second diameter conjugate to DAE. Now, if one can show that

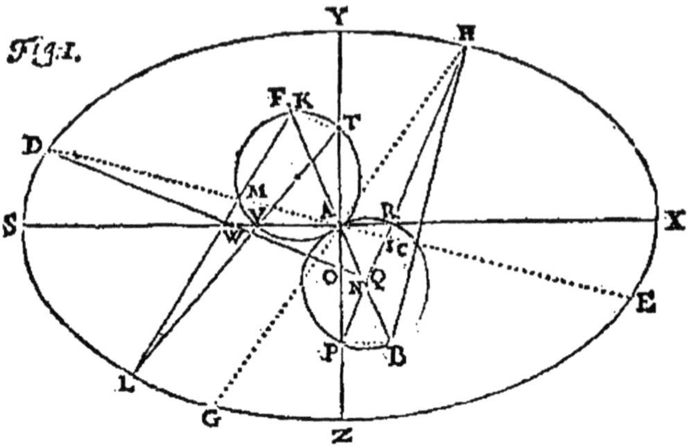

FIGURE 4.12

E^* coincides with $SYXZ$, then DAE and HAG are conjugate diameters of $SYXZ$ as well.

We prove the coinciding of $SXYZ$ and E^* as follows: We select an arbitrary point L on $SYXZ$ and we show that it also lies on E^*. To this end, a generator KM of E^* is selected in a way to be specified later; this generator is intersected by the generator TV of L (as a point on $SYXZ$). We shall denote this point of intersection by L' and we shall show that L' lies on E^* and also coincides with L.

Here Jan de Witt causes confusion as he also denotes this point of intersection L' by L.

In fact, he says (page [215]): "*Deinde juncta KM eaque producta versus L, agantur TK, PB ...*" : "Join first K and M, produce this line into the direction of (versus) L and then let TK and PB be drawn..." In the sequel of Jan de Witt's account, only the point L occurs; this is, however, sometimes the point L selected on $SYXZ$, sometimes the point of intersection of KM and TV, i.e., the point that we denoted by L',–as a provisional distinction from L– and that we prove to coincide with L.

In Figure 4.12a this situation is (exaggeratedly) represented. This kind of confusion occurs in more places, as mentioned before. Jan de Witt often draws a line through two points A and B and calls it ABC, although he only later defines the meaning of C.

We summarize the proof as follows and confine ourselves to Figure I on page [214]. Here OWD is the generator of D on the ellipse $SYXZ$.

First we rotate AOW counterclockwise "back" (*reciproce*) through 90°, so that we get ΔARP. Then we produce PR in the direction indicated and intersect the produced part with the ellipse $SYXZ$. This point of intersection is the point H, announced earlier in this annotation, and hence we

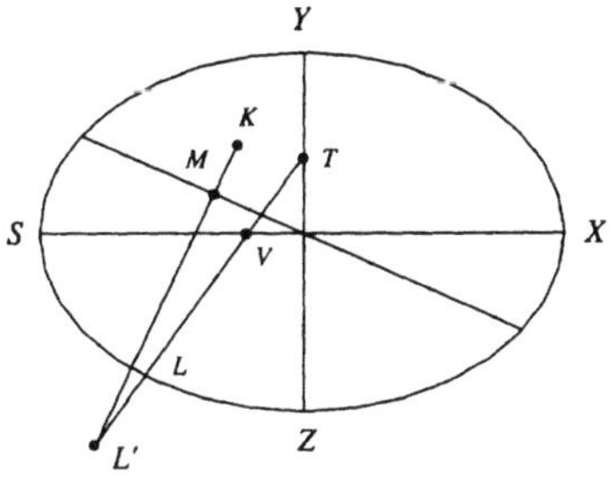

FIGURE 4.12a

have $PH = OD$. Finally, we draw HC perpendicular to DE and measure $HB = DA$ along HC.

Next we consider the ellipse E^* with DAE and HAG as diameters ($HA = AG$), which are evidently conjugate (i.e., transverse and second, respectively) diameters. In order to prove that $SYXZ$ coincides with E^*, we select an arbitrary point L on $SYXZ$ with generator TVL. The circle with diameter TV passes through A ($\angle TAV = 90°$), intersecting the produced part of BA at K and DA at M. We denote the point of intersection of KM and TV by L' and shall show:

a) L' lies on E^*; b) $L' = L$

and hence the point L lies on E^* too.

To this end, first TK and PB are drawn.

If Q is the point of intersection of DO and HP, then

$$\angle DQP = 90° = \angle HCI. \ (\Delta WAO \sim \Delta PQO!)$$

Also $\quad \angle DIQ = \angle HIC,$

hence $\quad \Delta DIQ \sim \Delta HIC$

and $\quad \angle ADO = \angle BHP.$

But $\quad OD = PH$ en $DA = HB,$

so $\quad \Delta ODA \simeq \Delta PHB,$

hence also, $\quad OA = PB.$

Furthermore, $\quad OA = AR,$

so $\quad PB = AR.$

The congruence

$$\Delta ODA \simeq \Delta PHB$$

implies $\angle HPB = \angle DOA = \angle WOA = \angle PRA,$

hence $PB = AR$ and $PB \parallel AR.$

Then it is immediately evident that $\Delta APB \simeq \Delta PAR$ and hence $AB = PR$ ($= TV$, as generator of the point L). This implies that the circle with diameter AB is congruent to the circle with diameter TV.

Furthermore, $\angle PBC = \angle TKM$ (via $\angle PAC = \angle TAM$),

hence $\angle PBH = \angle TKL',$ (*)

Also $\angle BPH = \angle BPR = \angle BAR$ (on arc BR)

and $\angle BAR = \angle KAV = \angle KTV = \angle KTL',$

so $\angle BPH = \angle KTL';$ (**)

furthermore, $\angle PAB = \angle TAK,$

hence the congruence of the circles implies

$$PB = TK. \quad (***)$$

(*), (**), and (***) then imply that $\Delta KTL' \simeq \Delta BPH$ and hence that $TL' = PH$.

From the construction of L it follows that $TL = PH$ and hence that L' coincides with L; therefore KM does intersect the generator TV of L at L. It remains to be proved that L' and hence L lie on E^*.

To this end, we remark that arc KM = arc BC, for $\angle KAM = \angle BAC$, hence–because of the congruence of the circles–$KM = BC$ and this is precisely the generator of E^*.

We also saw that $KL' = HB$ (because $\Delta KTL \simeq \Delta BPH$), hence L' lies on E^*; but because L' and L coincide, L lies on E^* as well.

Now L was an arbitrary point on E, hence E lies entirely on E^*, but then DE and HG are also conjugate diameters of the original ellipse $SYXZ$.

[3.16] The reasoning is the following: an ellipse E is given by the conjugate diameters DAE and HAG (see Figure I on page [219]). This implies that DAE is a transverse diameter and HAG is the secant belonging to it, which implies that the parts HC and HB of the line HCB (drawn perpendicular to DAE such that $HB = AD$) are the intervals of the ellipse E, that BC is the generator, and that BAC is the active angle.

Next –by a method to be specified later–two mutually perpendicular chords SAX and YAZ are constructed, which bisect each other at A. The essence

of the proof is that the ellipse that can be constructed with SAX and YAZ as axes coincides entirely with the original ellipse E, of which the axes SAX and YAZ have been constructed in this way.
This construction proceeds as follows. One draws HCB perpendicular to DAE such that $HB = AD$; as said earlier, HC and HB are the intervals. With AB as diameter, one draws a circle with center N. This circle intersects HN at P and at R.
The proposition is: if one measures the segments HP and HR along AR and AP, respectively, on both sides of A, then the line segments SAX and YAZ, thus obtained, are the axes of the given ellipse E.
The proof of this proposition proceeds as follows: One selects on AP the point O such that $AO = AR$.
Then one proves the congruence of the triangles DAO and HBP from the equations

i) $AO = BP$, since $AO = AR$, and $AR = BP$ because $\Delta APB \simeq \Delta PAR$.
ii) $\angle PBH = \angle PBC = \angle OAD$ since $\angle PBC + \angle PAC = 180°$ and $\angle OAD + \angle PAC = 180°$.
iii) $AD = HB$ (by construction).

This implies that $OD = PH$, but by construction $AS = PH$, hence also, $AS = OD$.
Furthermore, we show that $OW = RP$, for the triangles AOW and ARP are congruent, because

i) $AO = AR$;
ii) $\angle WAO = \angle PAR = 90°$;
iii) $\angle WOA = \angle PRA$ since $\angle WOA = \angle DOA = \angle HPB$ (this is implied by the congruence of the triangles DAO and HBP, which we have already proved) and since $\angle HPB = \angle RPB = \angle PRA$.

Summarizing, $OD = PH$, $OW = RP$, and hence $DW = RH$.
However, we selected AS and AY such that $AS = PH$ and $AY = RH$, so $OD = AS$ and $WD = AY$. Therefore the point D lies on the ellipse generated by the axes ASX and YAZ, and with OWD as its generator. Likewise the point H also lies on the ellipse E, with PRH as generator, but then the ellipse $SYXZ$ coincides entirely with the ellipse E, given by the conjugate diameters DAE and YAZ, whose axes SAX and YAZ are now constructed.

[3.17] In the proof of Theorem XIII, the given ellipse E is determined by the (mutually perpendicular) axes SAX and YAZ (see Figure I on page [219]). DAE is one of its diameters, hence for the time being only a line through the center. In the course of the proof, the ellipse E^*, with HB and HC as intervals and $\angle CAB$ as active angle, appears to coincide with the ellipse $SYXZ$. This means that for the points D and E on E^*, $DA = HB$ and

$AE = HB$, because DAE is a diameter of E^* as well; so also $DA = AE$.

[3.18] A transverse diameter can also be interpreted as a second diameter, where the original second diameter can be interpreted as a directrix. Hence, the parallel line in question is parallel to a directrix and touches the ellipse because of Corollary 2 of Theorem XII (= Proposition 13) on page [209].

[3.19] Here, *recta* clearly means line segment or chord. The case in which the line in question through P is ordinate-wise applied to the diameter through P has been disregarded, unless one wants to consider the point of contact as a chord with zero length. Jan de Witt showed that this segment has only two points in common with the ellipse, and apparently deduces from the figure that this line segment lies totally within the ellipse.

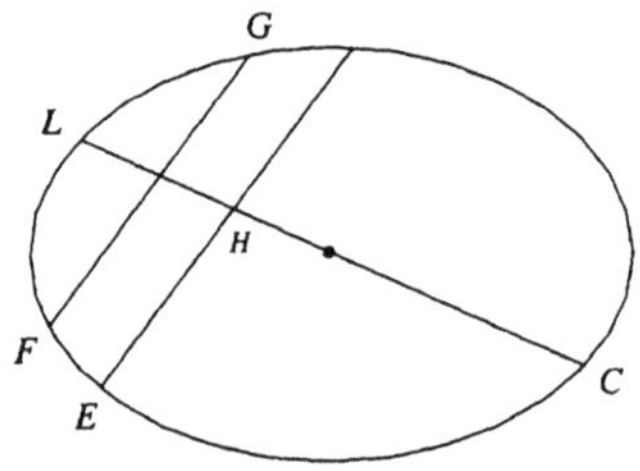

FIGURE 4.13

[3.20] The explanation is simple; see Figure 4.13. Here, $FI = IG$ and $EHL \parallel FIG$. Because of Theorem XIV (= Proposition 15), FIG is ordinate-wise applied to the diameter AKC and hence EHL–which is parallel to FIG–is ordinate-wise applied to AKC, but then according to Theorem XII, Corollary 4 (= Proposition 13) on page [211], $EH = HL$.

[3.21] Here the ellipse is seen as a given curve, on which one can select points without further construction.

[3.22] See Proposition 15, Corollary 2 (Theorem XIV).

[3.23] One first selects a pair of conjugate diameters according to Theorem XIV, Corollary 2 (= Proposition 15 on page [222]), and then the axes according to Theorem XIII, Corollary 1 (= Proposition 14 on page [218]).

[3.24] I.e., rotated counterclockwise through 90°; see page [214].

[3.25] Because of Theorem XIII, Corollary 4 (= Proposition 14) on page [220], the tangent to the ellipse at C is parallel to the diameter, conjugate to the diameter CF. The reasoning proceeds as follows: suppose ICK touches at C too, but differs from DCE. Then we draw through the center the lines $GH \parallel DCE$ and $LM \parallel ICK$; then: $ICK \neq DCE$, hence $LM \neq GH$. The diameters NO and CF conjugate to LM and GH are then different, hence $N \neq C$. If one draws $PNQ \parallel LM$, then PNQ touches the ellipse at N.
Because of Corollary 2 on pages [209]–[211], the ellipse (except N) lies entirely on one side of the tangent at N. Jan de Witt formulates this by saying that "C lies under PNQ." But then the line ICK lies entirely "under" the tangent PNQ as well, because ICK is parallel to PNQ and both lines

are parallel to LM. Since ICK was supposed to touch the ellipse at C as well, one can show –mutatis mutandis–in a similar way that N lies "under" ICK and hence PNQ lies entirely "under" ICK as well. These straight lines are, however, mutually different because $N \neq C$, and this results in a contradiction.

[3.26] For the Theorem in its most general form, we consider the figure on page [226]. Here we have

a) If DE touches the ellipse (with center A) at D and intersects the diameter IG at E and if CD is ordinate-wise applied to IAG, then $AE \cdot AC = AG^2$.

b) If A, D, E, and C are as above and if $AE \cdot AC = AG^2$, then DE touches the ellipse at D.

The proof is given in two steps: first, a) and b) are proved for the case in which the diameter in question is an axis (up to and including page [225], line 3 from the bottom); next the general case is proved, starting from the correctness of part b) of the "axes case."

i) In the first place, we direct our attention to the first figure on page [225]. Here AK and AG are the *semiaxes*. For the proof of a), we assume that ED touches the ellipse at D; OW is the generator of D, hence $OD = AG$ and $WD = AK$. The generator PR arises from OW by a counterclockwise (*in statione reciproca*) rotation through 90° and determines the point H on the ellipse. The angles at C, B, T, and F are right angles. Since AH and OD are mutually conjugate (see Problem II on page [222]) and as ED touches the ellipse at D, we have $ED \parallel AH$. The proof of proposition a) proceeds as follows:

$$\Delta OAW \simeq \Delta RAP \Rightarrow \Delta WCD \sim \Delta RTH \Rightarrow$$

$$\Rightarrow \Delta OBD \sim \Delta PFH.$$

However, $WD = RH$ (interval!) and $OD = PH$ (interval!), hence $WC = RT = FA$ and $DB = HF$.

Because $ED \parallel AH$ (see above) and $\angle C = \angle F = 90°$, we have $\Delta EDC \sim \Delta HAF$. This implies:

$$DC : AF = EC : HF.$$

We already saw $HF = DB$,

hence $\qquad DC : AF = EC : DB. \qquad (*)$

Because $\qquad \Delta DCW \sim \Delta DBO,$

we have: $\qquad DC : CW = DB : BO,$

but $$CW = AF,$$

hence $$DC : AF = DB : BO. \qquad (**)$$

(*) and (**) imply

$$EC : DB = DB : BO. \qquad (***)$$

In this case, one says that EC, DB, and BO are continuedly proportional. Next a property of proportions is used, for which de Witt refers to *Elementa* VI 20 and which one easily verifies:

$$a : b = b : c \Longleftrightarrow a : c = b^2 : c^2.$$

This implies for (***)

$$EC : BO = DB^2 : BO^2,$$

but $BO = CA$, hence

$$EC : CA = DB^2 : BO^2.$$

Via "*componendo*" (see Annotation [2.16]), one finds

$$(EC + CA) : CA = (DB^2 + BO^2) : BO^2,$$

i.e., $$EA : CA = DO^2 : BO^2.$$

Here it is essential that $\angle B$ is a right angle, i.e., we departed from the *axes* and not from arbitrary diameters!
However, $OD = GA$ and $BO = CA$, hence

$$EA : CA = GA^2 : CA^2.$$

One simply recognizes that (again see *Elementa* VI.20):

$$EA : GA = GA : CA$$

and hence $$AG^2 = AC \cdot AE. \qquad \text{q.e.d.}$$

The inverse of a) (i.e., proposition b) is trivial because of the uniqueness of the tangent, which was proved as Proposition 17.

ii) Next the case in which AG and AK are not semiaxes, but in which IG is an arbitray diameter (see the figure on page [226]).
In this figure, half the ellipse involved –to the right of IG–is represented as the arc $GDKI$. To prevent misunderstandings: the part of the figure to the left of IG is defined afterwards, but it is at any rate not the the second half

of the ellipse with a tangent! The proof is based on the correctness–proved

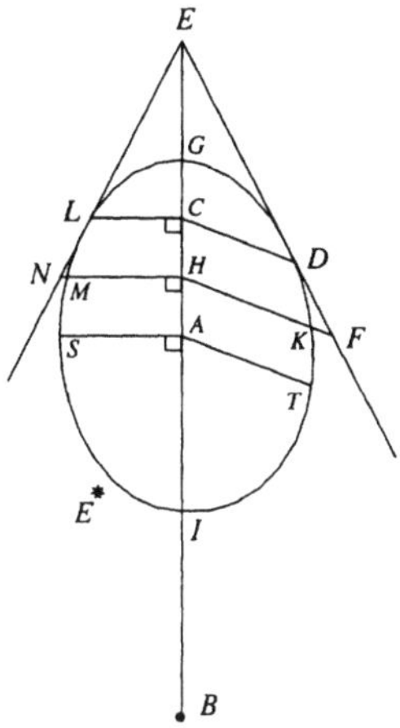

FIGURE 4.14

above–of part b) in the case of *axes*.
The segment CD is ordinate-wise applied to IG through the point D on the ellipse, hence is parallel to the diameter conjugate to IG. We denote the right half of this conjugate diameter by AT (T does not occur in the figure on page [226], but it does occur in Figure 4.14). E lies on the produced part of IG such that $AC \cdot AE = AG^2$. The proposition is that ED touches the ellipse at D. So here b) is proved first!

For the proof, we select on ED (or on its produced part) an arbitray point F ($F \neq D$) and we shall show that it does not lie on the ellipse. Consequently, ED is tangent to the ellipse. To this end, we construct a semiellipse E^* to the left of IG with IG as its axis and with the same parameter IB as the original ellipse (as a reminder: if a diameter has length $2a$ and its conjugate diameter has length $2b$, then the parameter belonging to it is $\frac{2b^2}{a}$. So here the parameter IB is equal to $\frac{2AT^2}{AG}$). At C, we raise the perpendicular to IG, which intersects E^* at L. Through F, we draw $HF \parallel CD$; the perpendicular at H on IG intersects E^* at M and EL at N. Because of what has been proved sub i (as part b of the proposition), it is clear that EL touches the ellipse E^* at L, because $LC \perp IG$, and so ordinate-wise applied to the axis IG, and according to the starting point, we have: $AG^2 = AC \cdot AE$. Apparently, part b) of the "axes case" holds in this case.
Next we first show that $LC = CD$ in the following way (for T see above): Because of Proposition 13 and because of the equality $IB = 2AT^2/AG$, we have:

$$\frac{DC^2}{GC \cdot CI} = \frac{AT^2}{AG^2} = \frac{IB}{2AG} = \frac{IB}{IG}. \qquad (*)$$

Because E^* has the same parameter IB, and hence $IB = \frac{2AS^2}{AG}$, we also have:

$$\frac{LC^2}{GC \cdot CI} = \frac{AS^2}{AG^2} = \frac{IB}{2AG} = \frac{IB}{IG}. \qquad (**)$$

Here AS is the semiaxis of E^*, conjugate to IG.
From (*) and (**), it appears immediately that $DC = LC$.
In a similar way one shows that $KH = MH$.
However, $CD : HF = EC : EH = CL : NH$, hence we also have

$$HF = HN.$$

Next, $NH > MH$ and because $NH = HF$ and $MH = HK$, we also have $HF > HK$, which proves that an arbitrary point ($\neq D$) on ED, or on its produced part, does not lie on the ellipse and hence ED touches the ellipse at D.
The converse of it (i.e., Proposition a) is again trivial, because of the uniqueness of the tangent at D.

[3.27] This means that F does not lie on the ellipse itself. By "ellipse," we mean the curve itself and not the area enclosed by it. One would almost come to the paradoxical translation "inside" for "*extra.*"

[3.28] See Proposition 16, Corollary on page [223].

[3.29] This is the quintessence of Jan de Witt's exposition "*absque ulla solidi consideratione,*" i.e., "without taking any solid object into consideration." In this context, see also the introductory letter.

[3.30] By "loci," we mean the sets of all points possessing a particular property. See also Annotation [1.20].
Compositio is translated here by "construction." Thus, Descartes says in his *Géométrie* (page 304) that he has treated some topics "*affin de faire voir qu"on peut construire tous les problèmes de la Géométrie ordinaire.*"

[3.31] For "genus" and "species" see Annotation [1.20].

[3.32] Presumably, Jan de Witt is thinking about an essay yet unwritten.

[3.33] See Annotation [1.20] and the introductory letter to van Schooten.

[3.34] This refers to *Liber Secundus*, which joins directly to the text translated here, and in which the curves discussed here are generated and treated by means of equations. Also, the equations in question are presented there in a standard form.

[3.35] The term "construct" has been selected as translation of "*Mechanica delineatio*" and indeed means drawing with the aid of a "mechanical" tool.

Chapter IV

[4.1] Jan de Witt speaks about the triangle ABC, but it is evident that, during the process, only A remains constant. If we take A as the origin of a rectangular system of axes, as in Figure 4.15, where GAF lies along the x-axis and EAD along the y-axis, then the point B slides along the y-axis and because the angle EBH remains 45°, the point C slides along the x-axis. So Figure 4.15 is supposed to present an arbitrary position. Here A is the fixed pole about which the line AK rotates, if the point K moves together with the point B, such that BK preserves a fixed length–say p. Now it is about the locus of the point of intersection of AK and the line through C parallel to the y-axis (say $O(x, y)$). One can simply verify this situation analytically. Figure 4.15 shows

$$\Delta ABK \sim \Delta OCA,$$

hence $$AB : BK = OC : CA;$$

but $$AB = AC = x,$$

so $$x : p = y : x$$

and hence $$x^2 = py.$$

This means that the locus of $O(x, y)$ lies on a parabola. Jan de Witt also

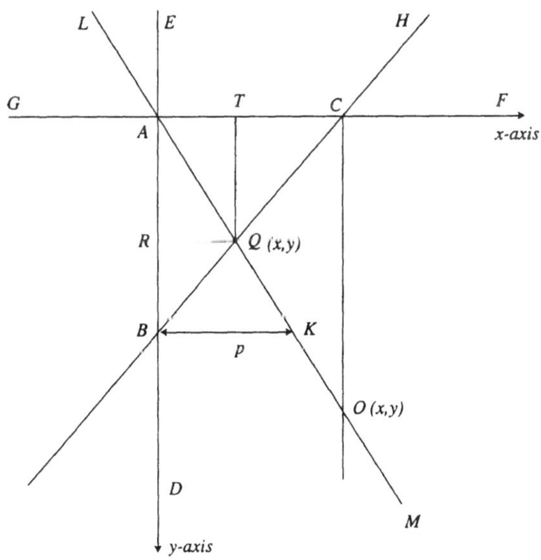

FIGURE 4.15

pays attention to the locus of the point of intersection $Q(x, y)$ of AK and BC.

From the similarity of ΔABK and ΔARQ, it follows that

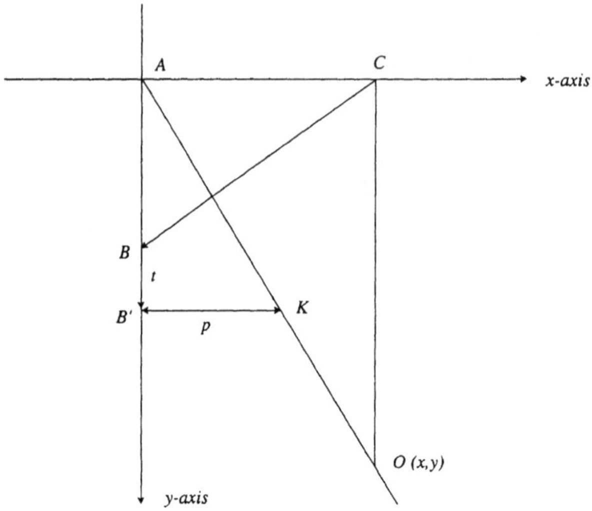

FIGURE 4.16

$$RQ : BK = AR : AB;$$

but $$AB = AR + RB = AR + RQ = y + x,$$

hence $$x : p = y : (y + x).$$

For the locus of $Q(x, y)$, we then have

$$x^2 + xy - py = 0.$$

Hence the locus of $Q(x, y)$ lies on a hyperbola.

Finally, Jan de Witt considers the case in which ΔABC is not isosceles, where the line segment of length $BK(= p)$ is not drawn from B, but from a different point–say B'–on the y-axis, which lies in our figure at a fixed distance t from the moving point B (Figure 4.16). In doing so, he only investigates the locus of $O(x, y)$. Then the following holds:

$$\Delta AB'K \sim \Delta OCA$$

implies $$AB' : B'K = OC : CA,$$

hence $$AB' : p = y : x.$$

However, $$AB' = AB + t.$$

If we represent the fixed ratio $AB : AC$ by m, then $AB = mx$, so

$$(mx + t) : p = y : x.$$

For the locus of $O(x, y)$, we have

$$mx^2 + tx - py = 0.$$

So this locus lies on a parabola as well.

To conclude, it may be remarked that it is evident here how loosely Jan de Witt handles the denotation of points; initially, he speaks about a line segment BC, "which is extended indefinitely on both sides, e.g., to D and E."

[4.2] Here, Jan de Witt anticipates the result that AD is the diameter of the parabola required.

[4.3] Here, Jan de Witt refers to Proposition 1 on page [162]. There a curve is generated with a given directrix, interval, and moving angle according to a specified process. In this case, the moving angle in the initial position equals the angle that the interval makes with the directrix (at the same side of the interval). This curve then appears to be a parabola. A self-evident question then is, What happens, if these angles are not equal? On page [178], Jan de Witt raises this question, but then says that, while he could treat it, he prefers a different approach for the next curve (the hyperbola), i.e., a rotating line pushing an angle instead of a rotating angle pushing a line. Now, on page [231], he returns to this topic and he shows that the case he had left unresolved on page [178] leads to a curve that has the characteristic property of a hyperbola–stated on page [181] and proved on page [182].

[4.4] The line of thought in the proof is the following: We confine ourselves to Figure I on page [232]. According to the prescription on page [159], here a curve is generated with active line IG (in the initial position), pole A, interval AL, and directrix KLO, but now with unequal angles IAL and KLA. The proposition is that a curve generated in this way is a hyperbola in the sense of page [182] (lines 10–13) and hence it possesses the characteristic property mentioned in Proposition 3 on pages [180] and [181]. In proving this, we select an arbitrary point D on the curve, generated by the moving angle in the position OAD, with O on the directrix and OD as its generator, i.e., parallel to AL.

Subsequently, the following construction is performed: First, the point K is defined on the directrix, such that the angle between KA and AL equals the angle between AL and the active line in the initial position. This line has already been denoted by AG, but the point G is only now specified in the following way: a circle is described, with center A and radius AK; this circle intersects the active line in the initial position at the points that are now denoted by I and G. This circle intersects the directrix at K and also at Q. The point of intersection of IK and GQ is denoted by F. The proposition is that the curve described in this way is a hyperbola in the sense of pages [180] and [181], with asymptotes FI and FG and with power (*potentia*, page [182]) $AB \cdot BF$. To this end, for an arbitrary point D on

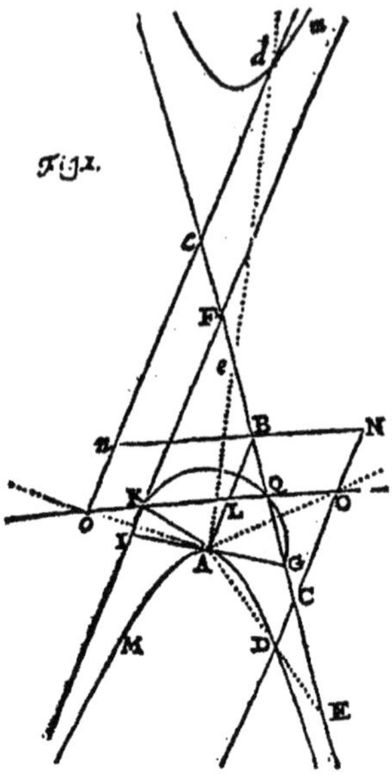

the curve, it is shown that $DC \cdot CF = AB \cdot BF$ ($CD \parallel IF$).

REMARK. This construction is based on the existence of the point K on the directrix; it fails if the angles between the interval AL and the directrix oLO and the active line IAG (in the initial position) are equal, as in the case of the parabola.

Then we would have: $\angle ALO = \angle LAG$; according to the construction, however, $\angle KAL = \angle LAG$, but then $\angle KAL = \angle ALO$ and hence $IAG \parallel oLO$, so the point of intersection K would not exist.

The proof now proceeds as follows: According to the construction,

$$\angle AIK = \angle AKI \text{ and } \angle KAL = \angle LAG = \tfrac{1}{2}\angle KAG.$$

For the exterior angle KAG of ΔKAI holds

$$\angle KAG = \angle AIK + \angle AKI,$$

hence $$\angle AIK = \angle GAL = \angle LAK,$$

so $$IF \parallel AL(\parallel OD).$$

Hence FG intersects AL and OD too, and $FB = BG$.

*

Furthermore, we remark that $\Delta LAK \sim \Delta OQC$,

for $\angle KLA = \angle QOC \quad (AL \parallel CO)$,

and $\angle LAK = \angle AIK \quad$ (see above),

and $\angle AIK = \angle OQC$,

because $IKQG$ is a cyclic quadrilateral.

If one also draws $BN \parallel QO$, then

$$\Delta KLA \sim \Delta COQ \sim \Delta CNB.$$

Since $IKQG$ is a cyclic quadrilateral, we also have

$$\angle AGC = \angle IKQ = \angle ALO.$$

Furthermore, $\angle LAO = \angle GAD$,

for $\angle LAO = \angle LAG - \angle OAG = \angle OAD - \angle OAG = \angle GAD$,

hence AD intersects QG (say at E) and

$$\Delta ALO \sim \Delta AGE,$$

so $AL : AG = LO : GE = BN : GE$, (1)

because $BN \parallel LO$ and $LB \parallel ON$.

**

From the previously proved similarity of ΔKLA and ΔCNB follows

$$AL : AK = BN : BC,$$

hence also $AL : AG = BN : BC$.

Together with (1), this implies

$$GE = BC.$$

But $GE = CE + GC$ and $BC = BG + GC$,

so $CE = BG (= FB)$.

But then $BE = GE + BG = BC + FB = FC$.

Finally, $\Delta ABE \sim \Delta DCE$,

hence $BE : CE = AB : DC$,

so $$DC \cdot BE = AB \cdot CE.$$

But $$CE = FB \text{ and } BE = CF,$$

hence $$DC \cdot CF = AB \cdot BF,$$

and this last product is constant, so the arbitrary point D on the curve has the characteristic property of an arbitrary point on the hyperbola with FG and FI as its asymptotes and $AB \cdot BF$ as its power (see Proposition 3 on page [180].)

[4.5] Until now, we assumed that in Figure I on page [232] the points I and K were different, like the points G and Q. The constructions in those cases in which I and K (and G and Q) are not different are treated on page [237].

In Figure III on page [233] and Figure II on page [236], I and K coincide and, consequently, FK touches the circle with center A and radius AK at K, which means that AK is perpendicular to FI. Also, it is immediately evident then that KO is perpendicular to FQG because $IKQG$ is a cyclic quadrilateral (see Figure I on page [232]), so $\angle KQB = \angle KIG$ and this remains true, as K approximates I (see also Figures I and II on page [236]). Furthermore, AL is perpendicular to KAG, because $\angle KAL = \angle LAG$ by virtue of the construction.

If the points Q and G in Figure I on page [232] coincide, then Figure IV on page [234] and Figure III on page [236] arise. In this case, FG touches the circle with center A and radius AK at G, and QK is perpendicular to FKI, because $\angle FKQ = \angle IGF$ (see again Figure I on page [232]). It is also clear that AL is perpendicular to KQ, because $AL \parallel IK$ by virtue of the construction and because now $IK \perp KQ$. If I and K coincide and G and Q coincide as well, the construction yields a parabola.

[4.6] In the first case, the asymptotes and the part of the tangent between the asymptotes are given. In the second case, two conjugate diameters are given in size and position. This case is immediately reduced to the first one (see Figure I on page [236]). Comparison with Figure I on page [232] then enables us to find the interval, the directrix, and the active angle of the hyperbola, determined by this data and thereafter to construct this curve point by point in line with the prescription on pages [156] and [160].

[4.7] See again Figure I on page [236]. The difference with the construction via a circular arc is minimal. Via parallel lines, one has to see to it that the angle FKO is assigned the correct value. In its practical implementation, however, this construction can be more accurate. Incidentally, it may be remarked that the point L is mentioned first (*ducta AL*), but defined last! Initially, AL is nothing more than a straight line parallel to the asymptote FI; then K and KO (also with undefined O) are constructed and only then L is defined as the point of intersection of KO and the line through A parallel to FI.

[4.8] In the construction just performed (Figure I on page [236]), it is assumed that neither the points K and I nor the points G and Q coincide.

The cases $K = I$ and $G = Q$ have been treated on page [235] and in Annotation [4.5]. Now the constructions belonging to these cases follow.

In the Figure on page [237], a hyperbola is generated by the asymptotes FS and FT and also by an arbitrary tangent ST. One now wants to construct a hyperbola such that either the active angles are right angles or the generator (in the initial position) is perpendicular to the directrix.

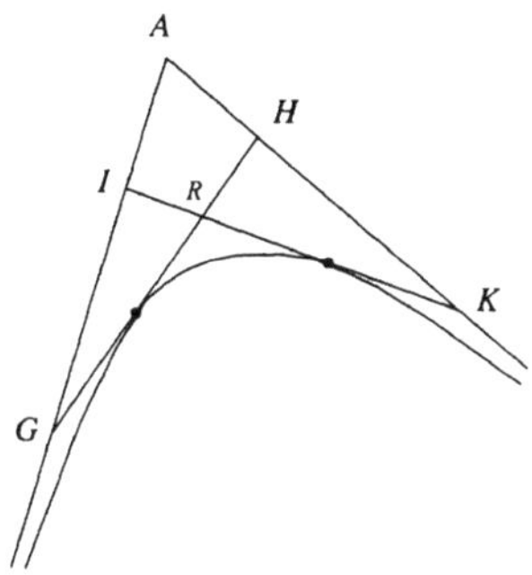

FIGURE 4.17

To this end, we go back to Proposition 9 on page [195]. This states the following: if AG and AK are the asymptotes of a hyperbola and if GH and IK are tangents to that hyperbola, intersecting at R (see Figure 4.17), then the following proportion holds:

$$AG : AI = AK : AH.$$

The converse also holds in a sense: if this proportion holds and, e.g., IK touches the hyperbola, then GH touches the hyperbola as well.

In the figure on page [237], one first raises the perpendicular TV to FT at T and then determines the point I on FV such that

$$FV : FI = FI : FS.$$

If we then draw a straight line IG through I, parallel to VT, then

$$FV : FI = FT : FG,$$

hence
$$FI : FS = FT : FG.$$

But then IG will touch this hyperbola too, by Proposition 9, because ST touches the hyperbola. Subsequently, one describes a semicircle on IG as diameter; then this semicircle touches the asymptote FT at G, because IG is perpendicular to FT. If one draws through the midpoint A of IG (the point of contact) a line AL perpendicular to KG, then a comparison with Figure III on page [236] shows that the hyperbola with active line IG, directrix KO, and interval AL is a hyperbola whose interval is perpendicular to the directrix.

The other hyperbola required is obtained by drawing AL perpendicular to IG and by making a comparison with Figure II on page [236].

[4.9] "Parallel to itself" means, as elsewhere in this work, "preserving its direction."

[4.10] In the first figure on page [238], we have

$$AB : BC = AL : LD, \text{ so } AB^2 : BC^2 = AL^2 : LD^2.$$

Now, $GAHFEG$ is the circle with center B and radius BA, hence

$$AB = GB; \qquad AL^2 = GL \cdot LH;$$

also,

$$BC = BK,$$

hence for an arbitrary point D on the locus holds

$$DL^2 : GL \cdot LH = DL^2 : AL^2 = BC^2 : AB^2 =$$

$$BK^2 : BG^2 = IK^2 : GH^2.$$

Therefore, this point satisfies the characteristic property of a point on the ellipse with GH and IK as its axes (Proposition 12, page [205]).

As in many other cases, Jan de Witt does not treat the question if the entire ellipse is traversed. For the origin of this construction, see section 8, ii of the Introduction.

A simple analytical proof is self-evident; to this end, assume a rectangular system of axes, take B at the origin O and let ABC revolve about O.

If $AB = a$ and $BC = b$, then from Figure 4.18 it is clear that for $D(x, y)$ holds

$$x = a\cos\alpha \quad \text{and} \quad y = -b\sin\alpha,$$

hence

$$\frac{x^2}{a^2} + \frac{y^2}{b^2} = 1.$$

[4.11] It concerns the line segment DB on the produced part of CB and the line segment BG on the produced part of AB. This is alluded to by "*nempe dictorum crurum, si opus fuerit, productorum.*"

[4.12] The problem is the following: ABC is an angle α, with arms BA and BC; the line l is perpendicular to BA at A and intersects the produced part of CB at D. Next, the angle ABC rotates about the pole B until it is in the position HBI (see figure on page [240] and Figure 4.19). Through H, one draws the line l' parallel to l and, through I, one draws the line m' parallel

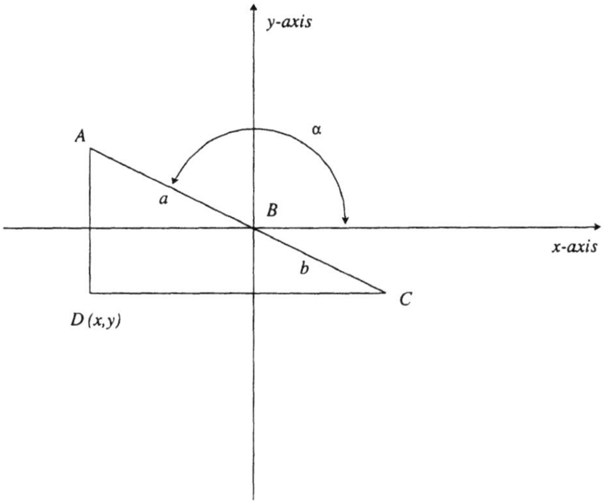

FIGURE 4.18

to BC; these lines intersect at K. Now we are interested in the locus of K when the angle ABC rotates about the pole B and when lines are drawn similar to l' and m'.

The proposition is that this locus is an ellipse, with center B, of which BD is a semidiameter. The conjugate (semi)diameter BP is parallel to l and has the length of BG; here G is the point of intersection of the produced part of AB and the perpendicular to BC at C (see figure on page [240]).

To clarify Jan de Witt's proof, we consider the figure on page [240] and Figure 4.19. Here we have:

$$AB = HB;\ BC = BI;\ AD \perp AB;\ HO \perp AB.$$

The line through I, parallel to BC, intersects the produced part of HO at K. IL and KM are perpendicular to DF. The perpendicular to BC at C intersects the produced part of AB at G. Besides, $BP = BG$ and $BP \parallel AD$.

First of all, we see that

$$\Delta HOB \sim \Delta ILB,$$

for $\qquad \angle OBL = \angle HBI,$

hence $\qquad \angle HBO = \angle IBL$

and furthermore

$$\angle HOB = \angle ILB = 90°.$$

Also, $$\Delta CBG \sim \Delta MKN,$$

for both triangles are similar to ΔOBN. This will be used later; first, we remark

$$DB^2 : NB^2 = AB^2 : OB^2 = HB^2 : OB^2.$$

Now, Jan de Witt applies "*conversio*" (see *Elementa* V, Definition 16); in this case:

$$a : b = c : d \Longleftrightarrow a : (a - b) = c : (c - d).$$

Here this yields

$$DB^2 : (DB^2 - NB^2) = HB^2 : (HB^2 - OB^2),$$

so $$DB^2 : DN \cdot NF = HB^2 : HO^2. \quad \text{(N.B.:} BF = DB)$$

Combined with the similarity of ΔHOB, ΔILB, ΔKMN, and ΔBCG, this implies

$$DB^2 : DN \cdot NF = HB^2 : HO^2 = BI^2 : IL^2 =$$

$$BC^2 : KM^2 = BG^2 : KN^2 = BP^2 : KN^2,$$

hence $$DB^2 : PB^2 = DN \cdot NF : KN^2.$$

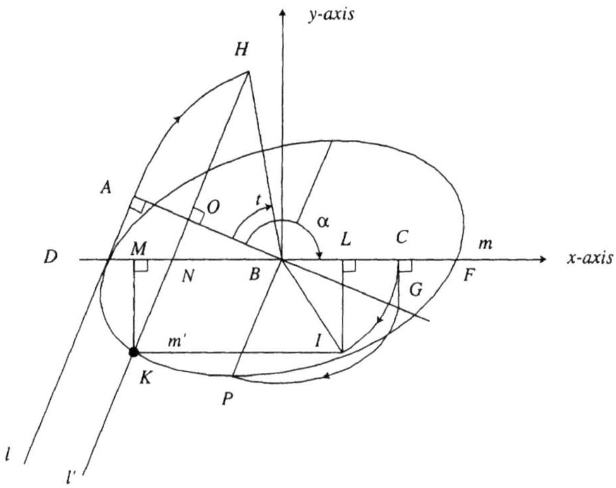

FIGURE 4.19

This means, however, that the point K has the characteristic property of the points on the ellipse with semiaxes DB and PB (Proposition 13, page [205]).

REMARK. If $\angle ABC$ is a right angle, then $AD \parallel BC$ and OH is in all positions of $\angle HBI$ parallel to BC. There are only two positions of $\angle HBI$ possible in which HK and KI intersect, but in those postions, they completely coincide (see Figure 4.20 and the following analytical derivation).

Finally, an analytical treatment of this problem. To this end, we assume the angle $ABC(=\alpha)$ to be located in a rectangular system of axes, such that B coincides with the origin and BC lies along the positive x-axis (see Figure 4.19).

Here again, $AB = a$ and $BC = b$. When ABC rotates through t° (clockwise) then the coordinates of $H(x,y)$ satisfy

$$x = a\cos(\alpha - t), \quad y = a\sin(\alpha - t).$$

For the point $I(x,y)$ we have

$$x = b\cos t, \qquad y = -b\sin t.$$

We assume $\alpha \neq 0^\circ$ and $\alpha \neq 180^\circ$, since $\alpha = 0^\circ$ and $\alpha = 180^\circ$ have been treated before.

The line through H perpendicular to AB has as its slope

$$m = \tan(\alpha - 90^\circ) = -\cot\alpha.$$

For the point of intersection $K(x,y)$ of the perpendicular to AB, through H, and the line through I, parallel to the x-axis, we therefore have

$$\left.\begin{aligned} y - a\sin(\alpha - t) &= -\{x - a\cos(\alpha - t)\}\cot\alpha, \\ y &= -b\sin t \end{aligned}\right\}$$

or

$$\left.\begin{aligned} y - a\sin\alpha\cdot\cos t + a\cos\alpha\cdot\sin t &= -x\cot\alpha + \\ &\quad + a\cot\alpha\cdot\cos\alpha\cdot\cos t + \\ &\quad + a\cot\alpha\cdot\sin\alpha\cdot\sin t, \\ y &= -b\sin t. \end{aligned}\right\} \quad (*)$$

After some reduction, this yields:

$$\begin{aligned} (y + x\cot\alpha)\sin\alpha &= a\cos t. \\ y &= -b\sin t. \end{aligned}$$

Elimination of t yields the equation for the locus of K:

$$\frac{(x\cos\alpha + y\sin\alpha)^2}{a^2} + \frac{y^2}{b^2} = 1.$$

It is immediately evident that for $\alpha = \pm 90°$,

$$\frac{y^2}{a^2} + \frac{y^2}{b^2} = 1$$

hence
$$y = \pm\frac{ab}{\sqrt{a^2+b^2}}\,.$$

Of course, this can also be deduced directly from (*). (Cf. also Figure 4.20.)

[4.13] For the term "*loci*," see Annotation [1.20].

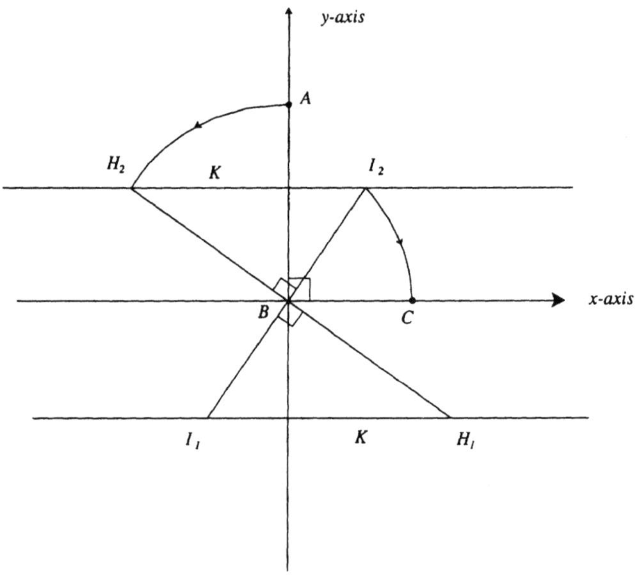

FIGURE 4.20

Appendix A
Application of areas

1. Jan de Witt associates the curves he introduces with the parabola, the hyperbola, and the ellipse, which were introduced by the Greeks, in six places in his book: on the pages [162], [164] (parabola), [182], [198] (hyperbola), and [208], [213] (ellipse).

His somewhat cryptical remarks refer to the names used by Apollonius of Perga (ca. 250–ca. 200 B.C.E) in his *Konika*. These terms were mentioned there for the first time in the history of mathematics (*Konika* **I**, 11, 12, and 13 conclude with the definition of parabola, hyperbola, and ellipse, respectively) and they refer to a characteristic property ($\sigma\acute{\upsilon}\mu\pi\tau\omega\mu\alpha$) of each of these curves. They are closely related to the so-called application problems treated by Euclid in his *Elementa* (I, 44, 45; II, 14; and VI, 28, 29).

In Appendices A and B, some topics that link the contributions by Euclid and by Apollonius have been put together for a better understanding of the allusions by Jan de Witt.

The application problems we encounter in Euclid"s *Elementa* seem to originate from the Pythagorean School, as–according to Proclus (410–485 C.E.)–was claimed by Eudemus (ca. 350 B.C.E.). In his commentary on the first book of the *Elementa*, Proclus writes as an explanation of Theorem I, 44 on the elliptic application:

> *According to the Eudemean School, these are ancient discoveries by the Muse of the Pythagoreans: the application of areas, their*

exceeding and their falling short. The more recent geometers derived the names from them and transferred these to the so-called conic curves and called one of them the parabola, one of them the hyperbola, and one of them the ellipse, as those ancient and godlike men saw that, what was signified by these names, in the application in the plane, of areas to a line segment.

2. Backgrounds of the application problems

a. The parabolic application.

Euclid poses the following problem in *Elementa* I, 44:

> *To a given straight line to apply, in a given rectilineal angle, a parallelogram equal to a given triangle.*

Without loss of generality, we will slightly modify this problem by replacing the given angle by a right angle and by replacing the given triangle by a square.

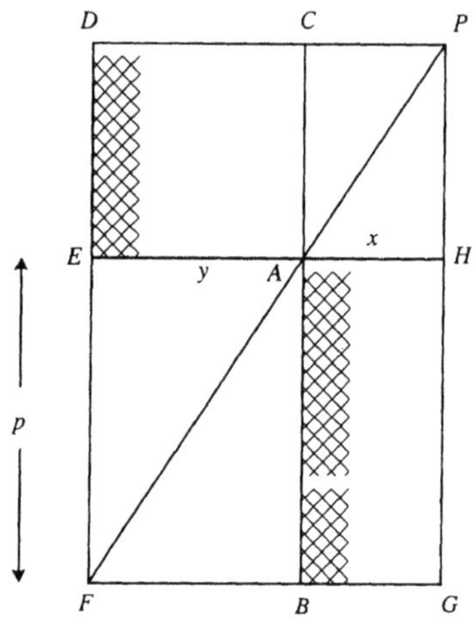

FIGURE A.1

In Figure A.1, let the line segment AB be the given line segment, having the length p, and let $ACDE$ be the given square, with y as the length of its sides; AB lies in the produced part of CA. The construction given here almost speaks for itself: first F, then FAP, etc.

Indeed: area $\Delta\ FPD$ = area $\Delta\ PFG$,

area $\Delta\ PCA$ = area ΔAHP,

area $\Delta\ AEF$ = area $\Delta\ FBA$,

hence area $ACDE$ = area $ABGH$.

If we also put $AH = x$, then we have

$$y^2 = px.$$

Later, Apollonius saw the link with the conic sections and derived the term "parabola" from this problem.

To us, it is clear from the last formula that the point $P(x, y)$ describes the upper half of a parabola when y varies. Reflection in the x-axis produces its other half. Figure A.2 has been added to establish the analogy with Figures A.3 and A.4. The term "parabolic application" has been derived from the

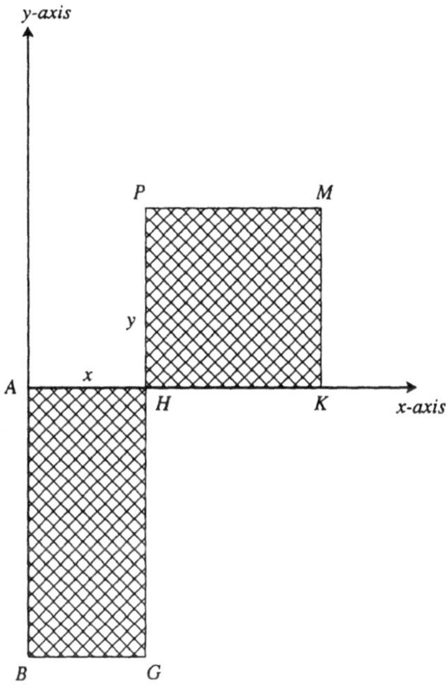

FIGURE A.2

word "παραβάλλειν" (*paraballein* = to compare, to apply: this accounts for our word parable, i.e., allegory). The connection with the concepts and the terminology of Apollonius will be given in Appendix B.

b. The elliptic application.
This problem was formulated by Euclid in *Elementa*. VI, 28 as follows:

> *To a given straight line to apply a parallelogram with the same area as a given rectilineal figure deficient by a parallelogrammic figure similar to a given one.*

The parallelogram that is missing will be called the defect.
Without further explanation Euclid added:

thus the given rectilineal figure must not be greater than the parallelogram described on the half of the straight line and similar to the defect.

We will see that this requirement is necessary for the existence of a solution. It can be interpreted as the requirement that the discriminant of the corresponding quadratic equation must be positive.

In this case too, we will modify the problem to some extent. First, we replace the given "rectilineal figure" by a square with the same area, which is possible by a suitable relocation of the vertices. At the same time we will consider a rectangle instead of the arbitrary parallelogrammic figure to which the defect is to be similar. Figure A.3 shows the required final situation. Here, AB is the given line segment to which the rectangle is to

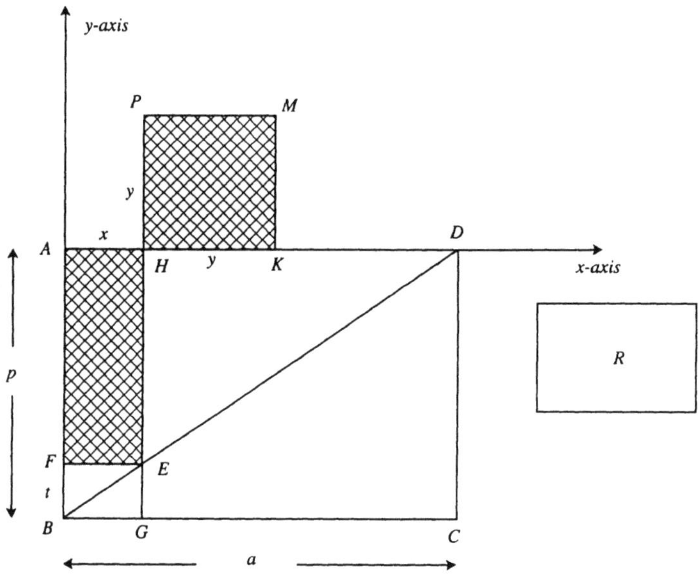

FIGURE A.3

be applied, $PHKM$ is the given square, and R is the given rectangle to which the defect is to be similar.

In order to obtain the final situation, we first construct the rectangle $ABCD$, similar to the given rectangle R and with one side equal to AB. The length of the other side BC –say a–is then fixed. Next we select on the diagonal DB a point E such that the area of the rectangle $AFEH$ equals the area of the square $PHKM$. The way in which this can be done is described in Section 3 of this appendix. It will then be evident that the defect $FBGE$ is similar to the given rectangle R as well.

In order to recognize the meaning of this construction, we take a rectangular coordinate system with A as its origin, and such that its positive

x-axis is along AD and its produced part, and its positive y-axis is along the produced part of BA.

If we put $AB = p$, $BF = t$, $AH = x$, and $HP = y$, then, because of the equality of the areas of $AFEH$ and $PHKM$,

$$y^2 = (p - t)x.$$

Furthermore, the similarity of $ABCD$ and $FBGE$ implies

$$a : x = p : t,$$

and hence

$$y^2 = (p - t)x = px - \frac{px^2}{a}. \qquad (*)$$

For Euclid, this construction of x means the solution of the quadratic equation in x (with fixed a, p, and y):

$$px^2 - apx + ay^2 = 0. \qquad (**)$$

The requirement that the discriminant D is nonnegative implies

$$a^2p^2 - 4apy^2 \geq 0, \text{ hence } y^2 \leq \frac{ap}{4},$$

which requirement Euclid–as has been said before–already mentioned while formulating the problem. Euclid had earlier solved the simultaneous system (*Elementa* II, 5 and 6)

$$\begin{aligned} x + y &= a, \\ xy &= b^2, \end{aligned}$$

in a geometric way. In fact, this yields the following quadratic equation in x:

$$x^2 - ax + b^2 = 0, \qquad (***)$$

with 1 as leading coefficient.

The construction we gave before leads to equation (**). Division by p yields (***). This operation is reflected in the choice of the rectangle R, whose sides are to each other as a is to p.

It is easy to verify that (**) corresponds to the equation

$$ay^2 = apx - px^2$$

of the ellipse with center $(a/2, 0)$, with axes of length a and $\sqrt{ap}$, respectively, and one vertex at the origin.

When the side y of the square $PHKM$ varies, the point E moves along the diagonal BD and the point $P(x, y)$ generates the upper half of the ellipse with the above equation. Reflection in the x-axis produces the lower half.

Later, Apollonius linked this problem to the ellipse as a conic section, and we will see in Appendix B how he came to this insight and also how he adapted his terminology.

Here too, a linguistic remark should be made: "ἐλλείπειν" (*elleipein*) is the Greek word for "to lack" or "to miss;" "ἔλλειψισ" (*elleipsis*) means "defect."

c. The hyperbolic application.
This problem has been formulated by Euclid in *Elementa* VI, 28 as follows:

> *To a given straight line to apply a parallelogram equal to a given rectilineal figure and exceeding by a parallelogrammic figure similar to a given one.*

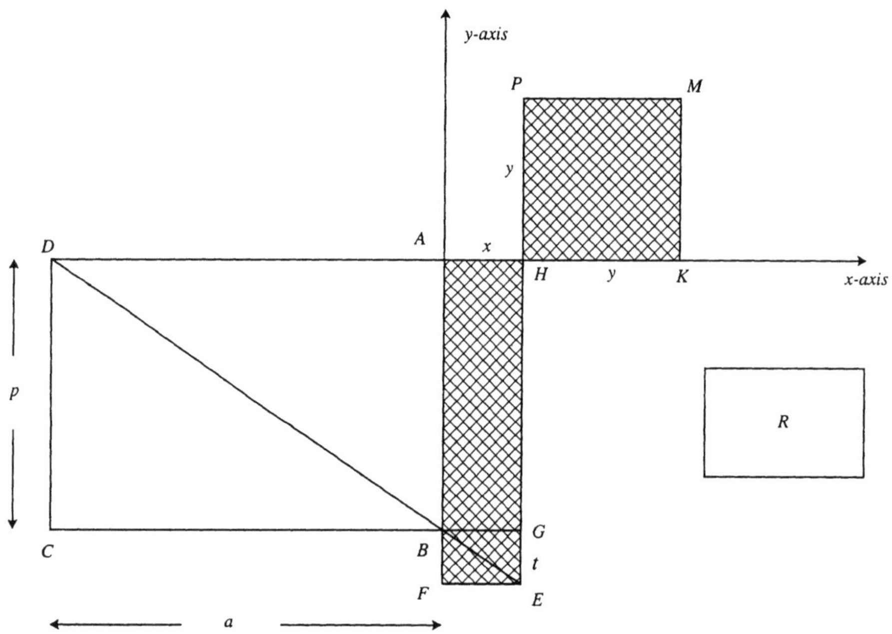

FIGURE A.4

The parallelogram that is exceeding will be called the "excess."

This problem is an analogue of the elliptic application, and we will modify it in the same way by replacing the given "recilineal figure" by a square with the same area and by considering a rectangle instead of the parallelogram to which the excess is to be similar.

In Figure A.4 a line segment AB is given, to which the rectangle is to be applied. $PHKM$ is the given square and R is the given rectangle to which the excess is to be similar.

In this case we require the construction of a rectangle $AFEH$ with one vertex at A, whose side AF lies along the support of AB, but now we require $AF > AB$ and the exceeding rectangle $BFEG$ has to be similar to the given rectangle R. Figure A.4 shows the final situation.

In order to obtain this situation, we first construct the rectangle $ABCD$, similar to the given rectangle R and with one side equal to AB. The length of the other side BC–say a–is then fixed. Next we select a point E on the produced part of the diagonal DB in such a way that the area of the rectangle $AFEH$ equals the area of the square $PHKM$. The way in which this can be achieved is described in section 3 of this appendix. It will be evident that the excess $BFEG$ is also similar to the given rectangle R.

In order to recognize the meaning of this construction, we use similar arguments as in the elliptic case. Here, too, we take a rectangular coordinate system with A as its origin, with the positive x-axis along DA and its produced part, and with its positive y-axis along the produced part of BA. If we put $AB = p$, $BF = t$, $AH = x$, and $HP = y$, then, because of the equality of the areas of $AFEH$ and $PHKM$, it follows that

$$y^2 = (p+t)x.$$

Due to the similarity of $ABCD$ and $GEFB$, we have

$$a : x = p : t,$$

and hence,

$$px^2 + apx - ay^2 = 0. \qquad (****)$$

Euclid had earlier solved the simultaneous system

$$\begin{aligned} y - x &= a, \\ xy &= b^2, \end{aligned}$$

(*Elementa* II, 5 and 6). In fact, this yields the quadratic equation with leading coefficient 1:

$$x^2 + ax - b^2 = 0.$$

The equation (****) is its generalization with leading coefficient $\neq 1$. Also in this case, dividing by p reflects the choice of the rectangle R whose sides have the ratio $p : a$.

If we compare (**) with (****), we see that of the four possible types of quadratic equations (with respect to the signs of the coefficients) two are missing, i.e.,

$$px^2 + apx + ay^2 = 0 \text{ and } px^2 - apx - ay^2 = 0.$$

The first one can be ignored since it has no positive solutions (a and p, being line segments, are of course positive!) and only positive solutions are accepted. The second equation can be reduced to (****) by the substitution $x = x' + a$.

In this way, all admissible quadratic equations have been solved constructively.

Just as in the preceding cases, an interpretation related to the conic sections is possible. If one writes (****) in the form

$$y^2 = px + \frac{px^2}{a},$$

then it is easy to verify that (****) corresponds to the equation of the hyperbola with center $(-a/2, 0)$, axes of length a and $\sqrt{ap}$, respectively, and one vertex at the origin.

When the side y of the square $PHKM$ varies, the point E moves along the produced part of the diagonal DB and the point $P(x, y)$ generates the upper half of the hyperbola with the above equation. Reflection in the x-axis produces its lower half.

Again a linguistic remark: "ὑπερβάλλειν" (*hyperballein*) is the Greek word for "to overshoot" and "ὑπερβολή" (*hyperbolè*) means "excess."

3. The constructive solution of the elliptic and hyperbolic application

a. As an introduction to the constructive solution of the elliptic and the hyperbolic application problems, we first give a tool: the construction of a rectangle, similar to a given rectangle R, and having a given area, which we can consider as the area S of a given square.

First we construct a rectangle H with the same area S, having one side in common with the given rectangle R (Figure A.5). We have seen before how this can be achieved by means of the parabolic application. Let us suppose that R has the sides k and m, and that H has the sides m and n. If we denote the sides of the required rectangle by x and y, then, because of the required similarity, the following holds:

$$x : y = k : m.$$

Because of the required equality of the areas,

$$xy = mn.$$

Elimination of y yields

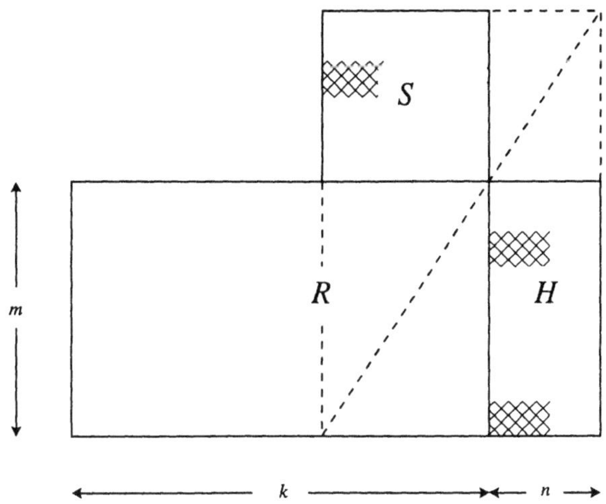

FIGURE A.5

$$x^2 = kn.$$

From this, x can be constructed in several ways and, applying the parabolic application, we can construct the remaining unknown y from $xy = mn$. Consequently, the required rectangle is known.

b. The constructive solution of the elliptic application.
The problem is (see Figure A.3) to construct a point E on the diagonal BD of the rectangle $ABCD$, such that the rectangle $AFEH$ has a given area y^2. To this end, we bisect AB at the point M (see Figure A.6). On MB

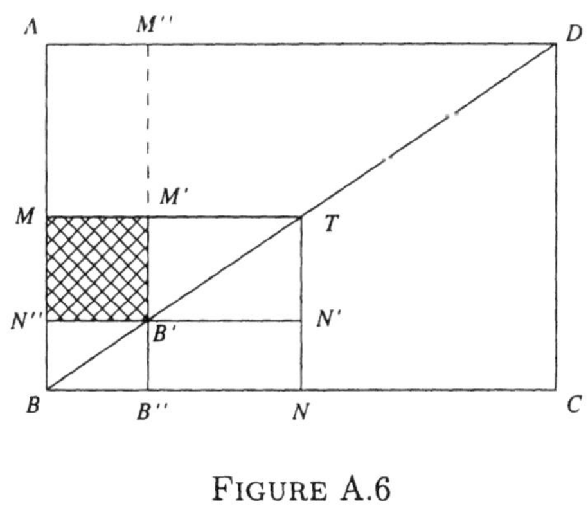

FIGURE A.6

we describe the rectangle $MBNT$, similar to $ABCD$ (hence T lies on the diagonal BD). As described in a, we construct in its "upper right angle" the rectangle $M'B'N'T$, similar to $MBNT$ and having an area, equal to the area of $MBNT$ minus y^2. Therefore, it is necessary that $y^2 \leq \frac{ap}{4}$, already required by Euclid when formulating the problem. The area of the "edge" or the "carpenter's square" (*gnomon*) $MBNN'B'M'M$ then has the given area y^2.

Next, we produce $B'M'$ and $N'B'$ till they intersect the sides AD, BC, and AB at M'', B'', and N'', respectively.

It is evident that area $B'B''NN'$ is equal to area $MN''B'M'$ (see the parabolic application), and it is also evident that area $AMM'M''$ is equal to area $MBB''M'$. If we "shift" $MBB''M'$ until it covers $AMM'M''$ and if we replace $MN''B'M'$ by $B'B''NN'$, then it is easy to see that the area of the edge (*gnomon*) $MBNN'B'M'M$ $(= y^2)$ equals the area of the rectangle $AN''B'M''$, applied to AB. This implies that the point B' coincides with the required point E on BD. We already mentioned that for the existence

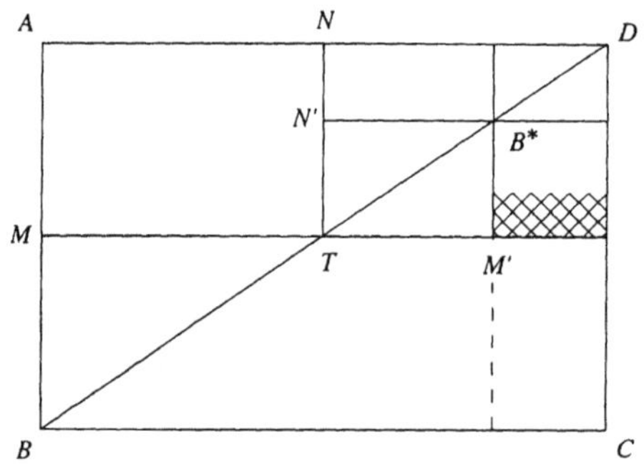

FIGURE A.7

of a solution it is required that

$$y^2 \leq \frac{ap}{4}.$$

As we have seen, this problem is equivalent to solving the following quadratic equation (in x):

$$px^2 - apx + ay^2 = 0.$$

So there are two positive solutions for x in this case, and consequently there are two possibilities for a constructive solution.

We find this second solution by placing the rectangle $TM'B'N'$ not "inside" the rectangle $TMBN$, as described in Figure A.6, but "outside." This results in the construction of Figure A.7, where the point B^* is a satisfactory solution as well. Euclid does not mention this solution. Several people have tried to explain this, all assuming that he did recognize the second possibility. Some people point out that he usually gives only one solution for a problem; others think that he delegated this second solution to his pupils as self-evident or too trivial.

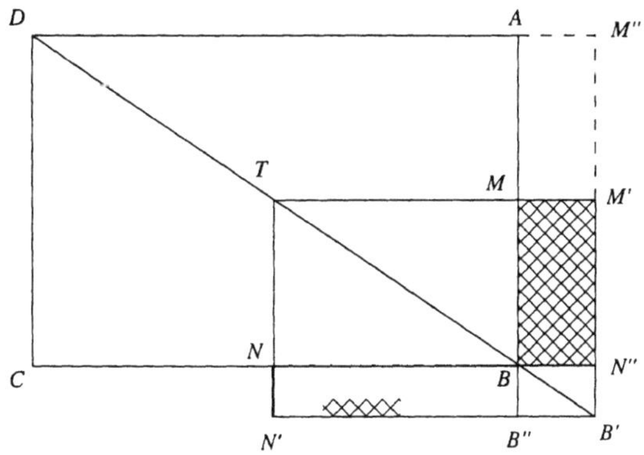

FIGURE A.8

c. The constructive solution of the hyperbolic application (Figure A.4). In this case the problem is to find a point E on the produced part of the diagonal DB of the rectangle $ABCD$, such that the rectangle $AFEH$ has a given area y^2. To this end, we again bisect AB by means of the point M (Figure A.8), and we describe on MB the rectangle $MBNT$ similar to $ABCD$ (so T lies on the diagonal BD).

Next, we construct the rectangle $TM'B'N'$ (similar to $TMBN$, and with an area equal to the area of $TMBN$ augmented by y^2) in the "upper left side" of $TMBN$. The *gnomon* $MM'B'N'NBM$ then has y^2 as its area. Finally, we produce $B'M'$ and DA till their point of intersection M'', and we produce CB until it intersects $B'M'$ at N''; likewise we produce MB till it intersects $N'B'$ at B''.

Now it is evident that the area of $NBB''N'$ is equal to the area of $MM'N''B$ and the area of $AM''M'M$ is equal to the area of $MM'N''B$. From this it is easily deduced that the area of the *gnomon* $MM'B'N'NBM$ $(= y^2)$ is equal to the area of the rectangle $AM''B'B''$, applied to AB. This implies that the point B' coincides with the required point E on the produced part of DB.

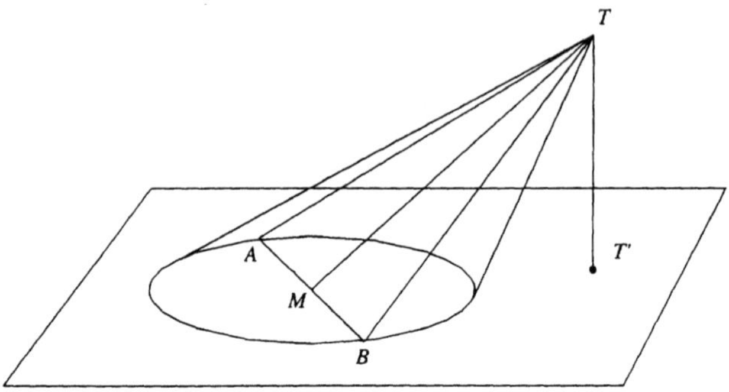

FIGURE B.1

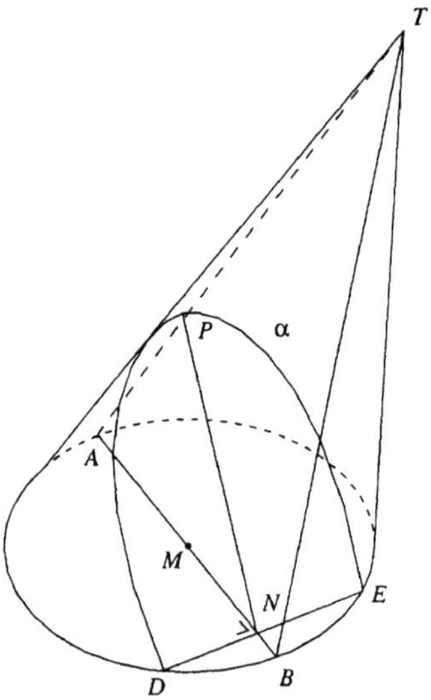

FIGURE B.2

Appendix B
The conic sections in Apollonius

a. In the first book of his *Konika*, Apollonius of Perga (ca. 250–ca. 190 B.C.E.) introduces our familiar conic sections as curves produced by the intersection of a plane with a ("double") oblique circular cone, i.e., a cone with a circle as its directrix and with a vertex whose projection on the plane of the circular base does not necessarily coincide with the center of this circle. He makes the usual distinction: the intersecting plane is parallel or not parallel to a generator and, if not, the intersecting plane cuts either one or two sheets of the cone.

First, Apollonius introduces the concepts *axis of a cone* and *axial triangle*. The axis of a cone is the straight line connecting its vertex with the center of the circular base.

An axial triangle is a triangle having the vertex of the cone as its vertex and a diameter of the circular base as its base, e.g., TAB in Figure B.1.

Suppose that a plane α intersects the cone, producing an intersecting curve (Figure B.2) and intersecting the plane of the circular base in the straight line DE. Then Apollonius selects an axial triangle (in this case TAB), the base AB of which is perpendicular to DE. Furthermore, we assume that the plane α intersects this axial triangle in the straight line PN. Apollonius proves that all chords parallel to DE are bisected by PN; for this reason PN is called a diameter. Nota bene: DNE is not necessarily perpendicular to the plane of TAB, and hence it is not necessarily perpendicular to NP, so Apollonius immediately proceeds very generally, without preference for principal (i.e., mutually perpendicular) axes.

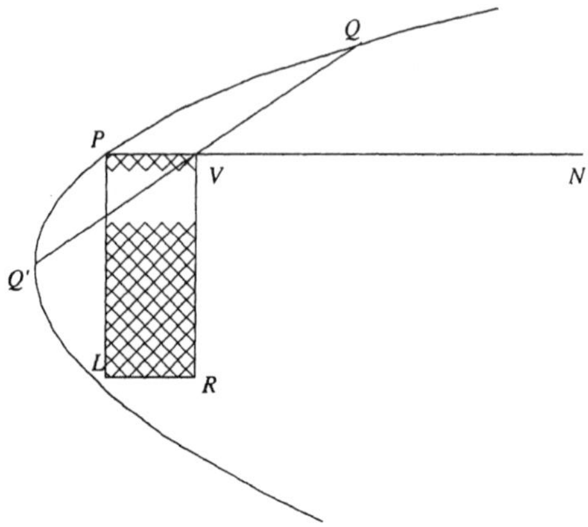

FIGURE B.3

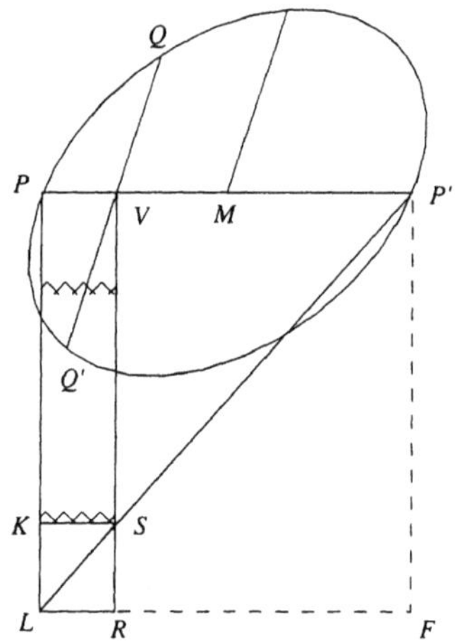

FIGURE B.4

Apollonius calls the straight lines parallel to DE "τεταγμένως καταγομέναι" (*tetagmenoos katagomenai*), which Jan de Witt translated verbatim by "*ordinatim applicatae*" and which we represent by "*ordinate-wise applied.*" These lines are parallel to the tangent to the curve at the corresponding vertex (vertices) of the diameter, i.e., the point(s) of intersection of this diameter and the curve.

A pivotal element in the reasoning of Apollonius is a special line segment PL, situated in the intersecting plane α and perpendicular to the diameter PN. Its length is selected in a particular and ingenious way and it depends only on the given cone, on the position of the plane of intersection and on the diameter PN of the curve of intersection (i.e., on the choice of the axial triangle).

We shall not discuss in detail how this length is determined. For this the reader is referred to the bibliography (e.g., [1], [23], [25]).

b. In the first case of intersection, i.e., the case in which the intersecting plane α is parallel to a generator, we consider the curve of intersection in its plane (see Figure B.3). Here, Q is an arbitrary point on this curve and QQ' has been ordinate-wise applied to the diameter PN (hence $QV = Q'V$). Apollonius proved that for this arbitrary point on the curve the following equality holds:

$$QV^2 = PV \cdot PL.$$

Here, PL is the line segment (independent of the choice of Q) that we mentioned before. This is clearly an instance of parabolic application, as has been discussed in Appendix A, since a rectangle $PLRV$ has been applied to PL with PL itself as a side, while the area of this rectangle equals the area of the square with QV as its side.

If we put $PL = p, PV = x$, and $QV = y$, then we see that

$$y^2 = px,$$

which we can interpret as the equation of a parabola, referred to a diameter and the tangent at its vertex as coordinate axes.

The link with the parabolic application was the reason for Apollonius to name this curve "παραβολή" (*parabolè* = parabola), which means "application" and to call the magnitude p "ὀρθία" (*orthia*), which we render as "erect side" or "latus rectum." Mydorge (1585–1647) introduced the term "parameter" for it.

c. The second case we consider is the situation in which the diameter of the curve of intersection meets one half of the cone at two points: P and P', i.e., the second case in which the plane of intersection meets only one half of the cone.

Figure B.4 shows a picture of this curve in its plane. The distance a between the two points of intersection P and P' plays an important role, together with the line segment PL, which, in this case too, has been appropriately chosen (see again [1], [23], [25]).

In Figure B.4, Q is an arbitrary point on the curve and QQ' has been ordinate-wise applied to the diameter PP', which here, too, means that QQ' is bisected by PP' (at V). The line segment PL is perpendicular to PP' and $P'L$ intersects the side RV of the rectangle $PVRL$ at S. Apollonius proved that for the arbitrary point Q on the curve the following equation holds:

$$QV^2 = \text{area } PLRV - \text{area } KLRS.$$

This is clearly an instance of elliptic application as treated in Appendix A: a rectangle $PKSV$ with the same area as the square with QV as its side has been applied to the line segment PL, while the "missing" rectangle, the defect, $KLRS$ is similar to the fixed rectangle $PLFP'$. "Fixed" here again means independent of the selection of the point Q on the curve, and solely depending on the cone, the selection of the plane of intersection, and the axial triangle, which in its turn determines the diameter PP'. The rectangle $KLRS$, moreover, has the "same position" as the rectangle $PLFP'$, which means that the corresponding sides are parallel.

If we put $PP' = a$, $PV = x$, and $QV = y$, then we conclude from Figure B.4:

$$\text{area } KLRS = \frac{x^2}{a^2}.\text{area } PLFP' = \frac{x^2}{a^2}.ap = \frac{px^2}{a},$$

and hence

$$y^2 = px - \frac{px^2}{a}.$$

For us, this is the equation of an ellipse, referred to a diameter and the tangent at one of its vertices as coordinate axes.

Apollonius adopted the term "ἔλλειψις" (*elleipsis* = ellipse) for this curve. In this case too, the terms that we mentioned before–ὀρθία, latus rectum, erect side, and parameter–are used for the line segment PL.

The segment PP' ($= a$) is called "πλαγία" (*plagia* = transverse side), "transverse diameter" or "*latus transversum.*"

d. In the remaining case, the plane of intersection α meets both halves of the cone and the intersecting curve consists of two branches. Again, we call its points of intersection with the chosen axial triangle P and P'. In this case, the support of PP' is diameter for the chords that have been ordinate-wise applied to PP', which means here too: parallel to the line in which the plane α intersects the plane of the circular base of the cone. These chords are parallel to the tangents at P and at P'. Figure B.5 displays the intersecting curve in its plane.

Here again, the point Q is an arbitrary point on the curve and QV has been ordinate-wise applied to the diameter PP' (hence $QV = Q'V$), while PL is again a line segment whose length has been appropriately selected, depending only on the cone, on the position of the intersecting plane, and on the chosen axial triangle that determines the diameter PP'.

PL is perpendicular to PP' and $P'L$ intersects the produced part of the side VR of the rectangle $PLRV$ at the point S.

Apollonius proved that, for the arbitrary point Q on the curve,

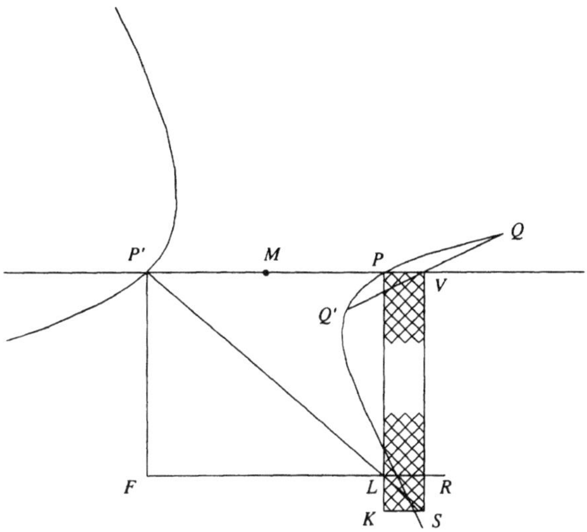

FIGURE B.5

$$QV^2 = \text{area } PLRV + \text{area } LKSR.$$

This is clearly an instance of hyperbolic application as treated in Appendix A: a rectangle $PKSV$, with the same area as the square with QV as its side, has been applied to the line segment PL, while the "exceeding" rectangle $LKSR$ is similar to the fixed rectangle $P'FLP$. The word "fixed" here has the same meaning as in sub c. Moreover, the rectangle $LKSR$ has the "same position" as the rectangle $P'FLP$.

If we put $PP' = a$, $PV = x$, and $QV = y$, then we conclude from Figure B.5:

$$\text{area } LKSR = \frac{x^2}{a^2} \cdot \text{area } P''FLP = \frac{x^2}{a^2} \cdot ap = \frac{px^2}{a},$$

hence $$y^2 = px + \frac{px^2}{a},$$

which we can interpret as the equation of a hyperbola, referred to a diameter and the tangent at one of its vertices as axes.

For this curve, Apollonius adopted the name "ὑπερβολή" (*hyperbolè* = hyperbola), which means "excess."

In this case the same terms as we mentioned before–ὀρθία, erect side, *latus rectum*, and parameter–are used for the line segment PL, and πλαγία (*plagia* = transverse side), transverse diameter, and *latus transversum* are used for the segment PP'.

e. In each case considered so far, one special diameter was involved, determined by the choice of the axial triangle. Immediately after this, Apollonius introduced other diameters. We will discuss the situation of each of the three cases: parabola, ellipse, and hyperbola.

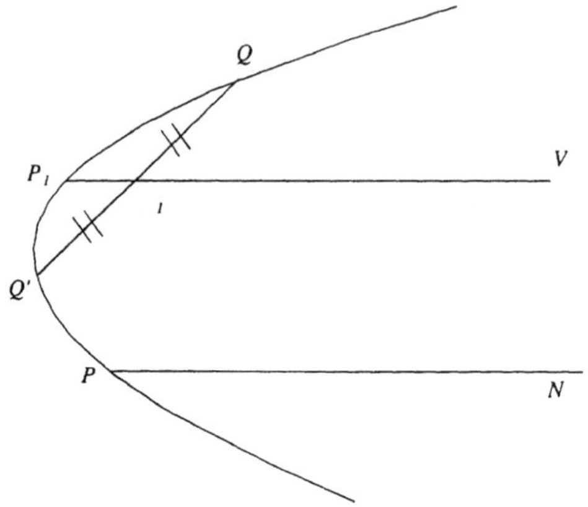

FIGURE B.6

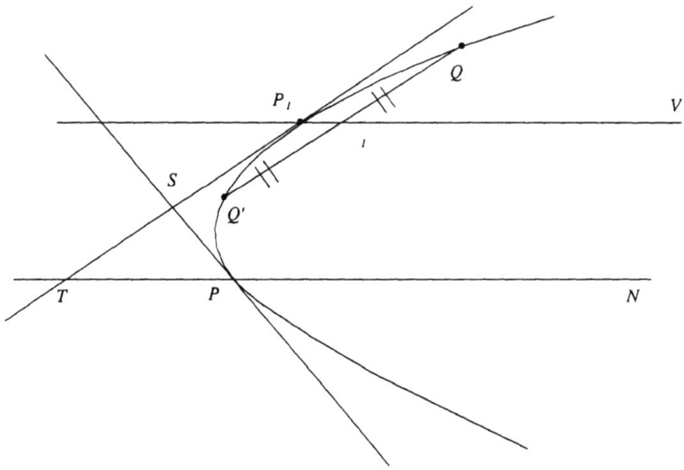

FIGURE B.7

i. The parabola.

For this case Apollonius proved that every straight line parallel to the diameter, as introduced before, is a diameter for the chords parallel to the tangent at the vertex of this straight line (i.e., the point of intersection with the parabola; see Figure B.6). These chords are also said to be ordinatewise applied to this diameter. To the diameter a parameter p' is assigned, which describes the parabola in the same way as in the case of the first diameter.

Apollonius derived the following relation for the new parameter p'(see Figure B.7)

$$P_1S : P_1T_1 = p' : 2P_1T. \qquad (*)$$

Here, T is the point of intersection of the tangent at P_1 and the (old) diameter PN; T_1 is the point of intersection of the tangent at P and the (new) diameter P_1N_1; and S is the point of intersection of P_1T and PT_1.

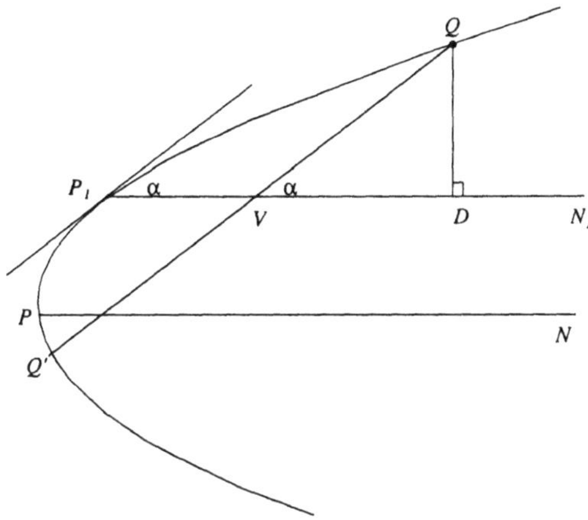

FIGURE B.8

If Q is an arbitrary point on the parabola and QQ' is ordinate-wise applied to the diameter P_1N_1, then the following relation can be derived by means of (*) (see [25]):

$$QV_1{}^2 - p'.P_1V_1,$$

or, if we put $P_1V_1 = x'$ and $QV_1 = y'$,

$$y'^2 = p'x',$$

which is formally the same equation as the equation with the "old" diameter deduced above.

Apollonius also introduced a particular diameter–the axis–i.e., the (unique) diameter perpendicular to the tangent at its vertex (see Figure B.8).

There is a relation between the parameter p_o corresponding to the axis and the parameter p belonging to an arbitrary diameter (see Figure B.8):

$$p : p_o = QV^2 : QD^2.$$

Here QQ' is ordinate-wise applied to the diameter P_1N_1, and is bisected

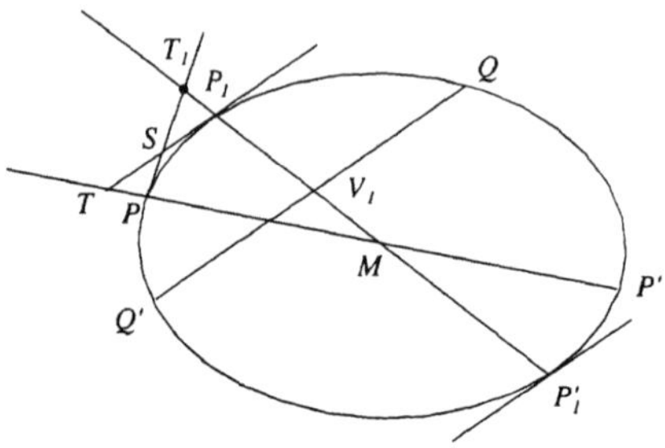

FIGURE B.9

by this diameter at V; QD is perpendicular to P_1N_1 and hence also perpendicular to PN. The ratio $QD : QV$ is the sine of the angle α between the tangent at P_1 and the axis of the parabola. The relation

$$p = \frac{p_o}{sin^2\alpha}$$

can be found in many textbooks of Analytic Geometry. See, e.g., [25].

ii. The ellipse.
For this case Apollonius proved that every chord passing through the center of the ellipse is a diameter for the chords parallel to the tangents at its vertices (see Figure B.9). These chords are also said to be ordinate-wise applied to this (new) diameter.

To this diameter a parameter p' is assigned, which describes the ellipse in the same way as in the case of the first diameter.

Apollonius derived the following relation for the new parameter p' (see Figure B.9):

$$P_1S : P_1T_1 = p' : 2P_1T. \qquad (**)$$

Here, again, T is the point of intersection of the tangent at P_1 and the diameter PP', T_1 is the point of intersection of the tangent at P and the diameter P_1P_1', and S is the point of intersection of PT_1 and P_1T.

Let Q be an arbitrary point on the ellipse and let QQ' be ordinate-wise applied to the diameter P_1P_1'. If we put $P_1V_1 = x'$, $QV_1 = y'$, and $P_1P_1' = a'$, then it can be proved by means of (**) that

$$y'^2 = p'x' - \frac{p'x'^2}{a'},$$

which is formally the same equation as the equation based on the "old" diameter, derived above.

If RR' is a chord passing through the midpoint of an arbitrary diameter PP' and if RR' is ordinate-wise applied to this diameter PP', then RR' is said to be conjugate to PP'. If one puts–as Jan de Witt does–$PP' = 2a$ and $RR' = 2b$, and if one denotes the parameter belonging to the diameter PP' by p, then it is easy to prove

$$p = \frac{2b^2}{a}.$$

iii. The hyperbola.
For this case Apollonius also proved that every chord passing through the center of the hyperbola is a diameter for the chords parallel to the tangents at its vertices (see Figure B.10). These chords are also said to be ordinate-wise applied to this (new) diameter. To this diameter a parameter p' is assigned, which describes the hyperbola in the same way as in the case of the first diameter.

Apollonius derived the following relation for the new parameter p' (see Figure B.10):

$$P_1S : P_1T_1 = p'' : 2P_1T. \qquad (***)$$

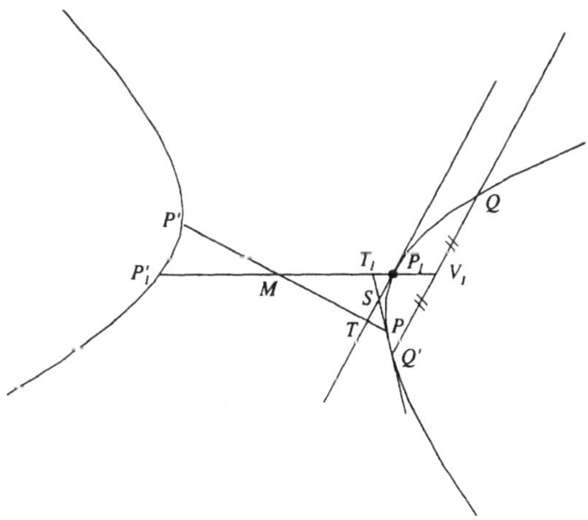

FIGURE B.10

Here, the points T, T_1, and S have the same meaning as in sub ii. Let Q be an arbitrary point on the hyperbola and let QQ' be ordinate-wise applied to the diameter P_1P_1'. If we put $P_1V_1 = x'$, $QV_1 = y'$, and $P_1P_1' = a'$, then it can be proved by means of (***) that

$$y''^2 = p''x'' + \frac{p''x''^2}{a''},$$

which is formally the same equation as the equation based on the "old" diameter, derived above.

In the case of a hyperbola, not all straight lines that have been ordinate-wise applied to a diameter (i.e., parallel to the tangents at its vertices) will intersect both branches of this curve. Nevertheless, for an arbitrary (transverse) diameter a conjugate diameter can be defined in the following way (see Figure B.11).

If PP' is an arbitrary diameter, intersecting both branches of the hyperbola, with length $2a$ and with corresponding parameter p, then we define the corresponding conjugate (second) diameter RR' as the line segment ordinate-wise applied to PP', passing through the midpoint M of PP' and having the length $2b$, where b is defined by

$$p = \frac{2b^2}{a}.$$

iv. The axes of an ellipse or a hyperbola are defined as the (unique) pair of diameters that are mutually perpendicular. We already encountered the axis of a parabola. We shall not go further into the many other related topics treated by Apollonius, as asymptotes, foci, and conjugate hyperbolas. For these we refer to the literature ([1], [23], [25]).

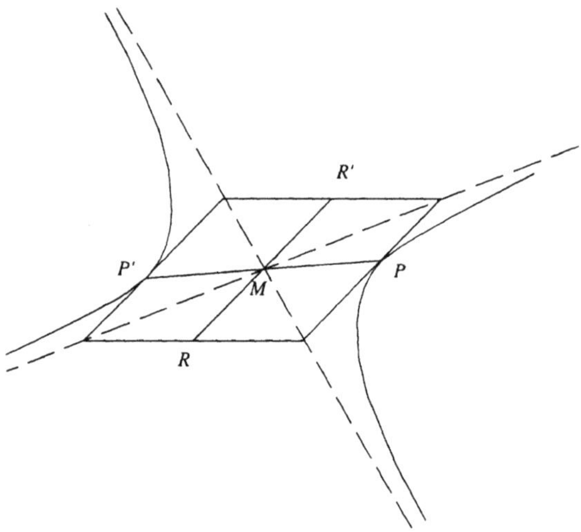

FIGURE B.11

References

[1] APOLLONIUS, *Apollonii Pergaei quae Graece exstant cum commentariis antiquis* edidit et Latine interpretatus est I. L. HEIBERG, Vols. I & II. Stuttgart, Teubner, 1974.

[2] ARCHIMEDES, *Opera Omnia cum Commentariis Eutocii* edidit I.L. HEIBERG, corrigenda adiecit E. S. STAMATIS. Stuttgart, Teubner, 1972.

[3] BOS, H.J.M., The Structure of Descartes's *Géométrie. Lectures in the History of Mathematics* **7**. Providence, Rhode Island, American Mathematical Society, 1997.

[4] BOYER, C.B., *History of Analytic Geometry*. New York, Scripta Mathematica, 1956.

[5] BUSARD, H.L.L., *Nicole Oresme, Quaestiones super Geometriam Euclidis*. Leiden, Brill, 1961.

[6] COOLIDGE, J.L., *The Mathematics of Great Amateurs*. New York, Dover Publications, 1963.

[7] COOLIDGE, J.L., *A History of the Conic Sections and Quadratic Surfaces*. New York, Dover Publications (reprint), 1968.

[8] COOLIDGE, J.L., *A History of Geometrical Methods*. New York, Dover Publications (reprint), 1963.

[9] DESCARTES, RENÉ, *Œuvres de Descartes*, ed. CH. ADAMS et P. TANNERY, Vol. VI, 1. Paris, Vrin/CNRS, 1964–1976.

[10] DESCARTES, RENÉ, *Œuvres Philosophiques*, ed. F. ALQUIÉ. Paris, Garnier Frères, 1963, 1967, 1973.

[11] DESCARTES, RENÉ, *The Geometry of René Descartes*, translated from the French and Latin by DAVID EUGENE SMITH and MARIA L. LATHAM. New York, Dover Publications, 1954.

[12] DESCARTES, RENÉ, *Geometrie*, Deutsch herausgegeben von Ludwig Schlesinger. Darmstadt, Wissenschaftliche Buchgesellschaft, 1969.

[13] DIJKSTERHUIS, E.J., *De Elementen van Euclides I & II*. Groningen, Noordhoff, 1930.

[14] DIJKSTERHUIS, E.J., *Archimedes I & II*. Groningen, Noordhoff, 1938. Translated into English by C. DIKSHOORN. Copenhagen, Ejnaar Munksgaard, 1956.

[15] EUCLIDES, *Euclidis Elementa*, post I.L. HEIBERG edidit E.S. STAMATIS, Vols. I–V. Leipzig, Teubner, 1969–1977.

[16] FLADT, K., *Geschichte und Theorie der Kegelschnitte und der Flächen 2. Grades*. Stuttgart, Ernst Klett Verlag, 1963.

[17] FRUIN, R., *Brieven van Johan de Witt*, edited by R. FRUIN and published by G.W. KERNKAMP and N. JAPIKSE. Vol. IV, Amsterdam, Johannes Müller, 1913.

[18] GEER, P. VAN, Johan de Witt als wiskundige. *Nieuw Archief voor Wiskunde* (2), XI, (1915), pp. 98–126.

[19] GILLESPIE, C. C. (ed.), *Dictionary of Scientific Biography* (VIII Vols.). New York, Charles Scribner's Sons, 1981.

[20] GROOTENDORST, A.W., Johan Hudde's Epistola Secunda de Maximis et Minimis. *Nieuw Archief voor Wiskunde* (4), V, (1987), pp. 303–334.

[21] GROOTENDORST, A.W., & J.A. VAN MAANEN, van Heuraet's Letter (1659) on the Rectification of Curves, *Nieuw Archief voor Wiskunde* (3), XXX, (1982), pp. 95–113.

[22] HEATH, SIR TH.L., *The Thirteen Books of Euclid's Elements*, translated from the text of Heiberg, with Introduction and Commentary by SIR THOMAS L. HEATH, Vols. I–III. New York, Dover Publications, 1956.

[23] HEATH, SIR TH.L., *Apollonius of Perga, Treatise on Conic Sections*. Cambridge, Barnes and Noble (reprint), 1961.

[24] HEATH, SIR TH.L., *The Works of Archimedes.* New York, Dover Publications (reprint), 1953.

[25] HEATH, SIR TH.L., *A History of Greek Mathematics*, Vols. I & II. Oxford, London, etc., Oxford University Press, 1965.

[26] HOFMANN, J.E., *Frans van Schooten der Jüngere.* Wiesbaden, Franz Steiner Verlag, 1962.

[27] HOGENDIJK, J.P., Book I of Jan de Witt's Elementa Curvarum Linearum and the Conics of Apollonius. *Nieuw Archief voor Wiskunde* (4), XVII, (1999), pp. 453–463.

[28] HUYGENS, CHRISTIAAN, *Œuvres complètes de Christiaan Huygens*, publiées par la Société Hollandaise des Sciences, Vols. I–XXII. Den Haag, 1888–1950.

[29] JAPIKSE, N., *Johan de Witt.* Amsterdam, Meulenhoff, 1915.

[30] KLINE, M., *Mathematical Thought from Ancient to Modern Times.* New York, Oxford University Press, 1972.

[31] KNORR, W.R., *The Evolution of the Euclidean Elements.* Dordrecht, Reidel, 1975.

[32] MAANEN, J.A. VAN, *Facets of Seventeenth Century Mathematics in the Netherlands.* Utrecht, Elinkwijk, 1987.

[33] MORROW, G., *Proclus; A Commentary on the First Book of Euclid's Elements.* Princeton, New Jersey, Princeton University Press, 1970.

[34] ROWEN, H.H., *John de Witt, Grand Pensionary of Holland, 1625–1672.* Princeton, New Jersey, Princeton University Press, 1978

[35] ROWEN, H.H., *John de Witt, Statesman of the True Freedom.* Cambridge, Cambridge University Press, 1986.

[36] SCHOOTEN JR., F. VAN, *Geometria à Renato Des Cartes Anno 1637 Gallicè edita; postea autem unà cum Notis Florimondi de Beaune, (...) in Latinam linguam versa, & Commentariis illustrata, opera atque studio Francisci à Schooten.* Leiden, Jan Maire, 1649. Revised and supplemented edition in 1659.

[37] SCOTT, J.F., *The Mathematical Work of John Wallis, D.D., F.R.S. (1616–1670).* London, Taylor and Francis, 1938.

[38] SCOTT, J.F., *The Scientific Work of René Descartes.* London, Taylor and Francis, 1976.

[39] STECK, M., *Proklos Diadochus (410–485); Kommentar zum Ersten Buch von Euklids "Elementen."* Halle, Deutsche Akademie der Naturforscher, 1945.

[40] STEINER, JACOB, *Geometrical Constructions with a Ruler* (a translation of his 1833 book). New York, Scripta Mathematica, 1950.

[41] THAER, C., *Euklid, die Elemente.* Buch I–XIII. Herausgegeben und ins Deutsche Übersetzt von Clemens Thaer. Darmstadt, Wissenschaftliche Buchgesellschaft, 1962.

[42] VER EECKE, PAUL, *Les Coniques d'Apollonius de Perge.* Œuvres traduites pour la première fois du Grec en Français. Bruges, Desclée de Brouwer et Cie, 1923.

[43] VRIES, HK. DE, Historische Studiën II, *N.T.v.W.* 12e Jrg. 1924/25, pp. 260–288. (here *Die geometrischen Constructionen* by Steiner[40] are annotated). Also published as chapter II in *Historische Studiën deel I.* Groningen-Batavia, Noordhoff, 1926.

[44] WAERDEN, B.L. VAN DER, *Science Awakening.* Groningen, Noordhoff, 1954.

[45] WALLIS, JOHN, *Opera Mathematica,* Vols. I–III. Oxford, Th. Robinson, 1697–1699.

[46] WALLIS, JOHN, *Opera Mathematica,* Vols. I–III. Reprint introduced by C.J. SCRIBA. Hildesheim-New York, Georg Olms Verlagsbuchhandlung, 1972.

[47] WITT, JAN DE, *Elementa Curvarum Linearum–Liber Primus.* Text, translation, introduction, and commentary (in Dutch) by A.W. Grootendorst. Amsterdam, Centrum voor Wiskunde en Informatica, 1997.

[48] WICKEFOORT CROMMELIN, H.S.M. VAN, *Johan de Witt en zijn Tijd.* Amsterdam, Nederlandse Algemene Maatschappij van Levensverzekering "Conservatrix," 1913.

[49] ZEUTHEN H., *Die Lehre von den Kegelschnitten im Altertum.* Hildesheim, Georg Olms Verlagsbuchhandlung, 1966.

Index

Sources and Studies in the History of Mathematics and Physical Sciences

Continued from page ii

E. Kheirandish
The Arabic Version of Euclid's *Optics*, Volumes I and II

J. Lützen
Joseph Liouville 1809–1882: Master of Pure and Applied Mathematics

J. Lützen
The Prehistory of the Theory of Distributions

G.H. Moore
Zermelo's Axiom of Choice

O. Neugebauer
A History of Ancient Mathematical Astronomy

O. Neugebauer
Astronomical Cuneiform Texts

F.J. Ragep
Naṣīr al-Dīn al-Ṭūsī's Memoir on Astronomy
(al-Tadhkira fī ᶜilm al-hay'a)

B.A. Rosenfeld
A History of Non-Euclidean Geometry

J. Sesiano
Books IV to VII of Diophantus' *Arithmetica*: In the Arabic Translation Attributed to Qusṭā, ibn Lūqā

B. Stephenson
Kepler's Physical Astronomy

N.M. Swerdlow/O. Neugebauer
Mathematical Astronomy in Copernicus's De Revolutionibus

G.J. Toomer (Ed.)
Appolonius *Conics* Books V to VII: The Arabic Translation of the Lost Greek Original in the Version of the Banū Mūsā, Edited, with English Translation and Commentary by G.J. Toomer

G.J. Toomer (Ed.)
Diocles on Burning Mirrors: The Arabic Translation of the Lost Greek Original, Edited, with English Translation and Commentary by G.J. Toomer

C. Truesdell
The Tragicomical History of Thermodynamics, 1822–1854

K. von Meyenn, A. Hermann, V.F. Weisskopf (Eds.)
Wolfgang Pauli: Scientific Correspondence II: 1930–1939

Sources and Studies in the History of Mathematics and Physical Sciences

Continued from the previous page

K. von Meyenn (Ed.)
Wolfgang Pauli: Scientific Correspondence III: 1940–1949

K. von Meyenn (Ed.)
Wolfgang Pauli: Scientific Correspondence IV, Part I: 1950–1952

K. von Meyenn (Ed.)
Wolfgang Pauli: Scientific Correspondence IV, Part II: 1953–1954

MIX
Papier aus verantwortungsvollen Quellen
Paper from responsible sources
FSC® C105338

If you have any concerns about our products,
you can contact us on
ProductSafety@springernature.com

In case Publisher is established outside the EU,
the EU authorized representative is:
Springer Nature Customer Service Center GmbH
Europaplatz 3, 69115 Heidelberg, Germany

Printed by Libri Plureos GmbH
in Hamburg, Germany